Signs of War and Peace

# Signs of War and Peace

Social Conflict and the Use of Public Symbols
in Northern Ireland

*Jack Santino*

palgrave

SIGNS OF WAR AND PEACE
© Jack Santino, 2001

First published 2001 by
PALGRAVE
175 Fifth Avenue, New York, N.Y.10010 and
Houndmills, Basingstoke, Hampshire RG21 6XS.
Companies and representatives throughout the world

PALGRAVE is the new global publishing imprint of St. Martin's Press LLC Scholarly and Reference Division and Palgrave Publishers Ltd (formerly Macmillan Press Ltd).

ISBN 0-312-23640-9 hardback

Library of Congress Cataloging-in-Publication Data

Santino, Jack.
    Signs of war and peace: social conflict and the use of public symbols in Northern Ireland/Jack Santino.
       p. cm. Includes bibliographical references and index.
    ISBN 0-312-23640-9
       1. Northern Ireland—History. 2. Signs and symbols—Northern Ireland—History— 20th century. 3. Social conflict—Northern Ireland—History—20th century. 4. Popular culture—Northern Ireland—History—20th century. 5. Public opinion— Northern Ireland—History—20th century. 6. Northern Ireland—Social conditions—1969– I. Title.

DA990 .U46 S27 2001
941.60824—dc21                                                                          00-069216

A catalogue record for this book is available
from the British Library.

Design by Newgen Imaging Systems (p) Ltd, Chennai, India.

First edition: July, 2001
10  9  8  7  6  5  4  3  2  1

Printed in the United States of America.

For Lorraine Lawrence and Eileen McManus,
and their families

# Contents

# Preface

*S*ingns of War and Peace focuses on the role that public display plays in the conflict in Northern Ireland. In so doing, it ranges freely over other times, places, and events that shed light on the social and political processes and dynamics involved in public-display traditions such as the Saint Patrick's Day parades in Boston and the popular spontaneous shrines to Lady Diana in London, following her death in Paris in 1997. The book is about the nature of public display and its relationships to class-based aesthetics, tradition, and popular style. It is also about contest, conflict, and civil war, and the ways in which they are all intimately intertwined, both in Northern Ireland and throughout the world.

I take as central what Beverly Stoeltje (1993) refers to as the "ritual genres"—here including parades, bonfires, effigy burnings, spontaneous shrines marking the sites of untimely deaths, processions, and demonstrations, along with components of these such as flags, banners, and murals, which are part of the festival context and also have lives of their own. I do not see these as peripheral to the well-known conflict, or as reflections of it. I see them as part of it, constitutive of the violence on some levels but oppositional to the violence on others. This is because the events and artifacts themselves involve members of society from different socioeconomic levels, religious backgrounds, and regions. The various ritual events and public artifacts represent different positions on some of the central questions in Northern Irish society. In this age of globalization, they are attempts to negotiate the local with the international; the (formerly) imperial with the national. As such, this study has a relevance wider than Northern Ireland itself because similar (though by no means identical) situations are occurring throughout the world with greater and greater frequency.

Despite the repeated emphasis on violence in the international news media and in scholarly books such as this one, Northern Ireland is—even before the cease-fires and the Good Friday agreement of 1998—largely a peaceful place. Statistically, violence there is much less frequent than in most large cities in the United States. Almost all the violence that does exist, even the most horrific, is carried out in the name of principle. My purpose in writing this book is to place the study of ritual, festival, holidays, celebrations, and public display in the center of the study of major social problems such as war, conflict, and violence. In doing so, I do not mean to further stereotype the much maligned six counties of the province of Ulster known as Northern Ireland. I love Ulster, and I firmly believe that the quality of life there is superior to that of the United States in many ways. I have found people of Northern Ireland of all backgrounds, political points of view, and religions to be unfailingly generous, cheerful, engaged, and intelligent—a joy to be with. They have lived in an untenable situation for decades, but the possibility for a lasting peace is now at hand. This book is for all of them, but I want to single out Eileen McManus and Lorraine Lawrence—both of whom know the suffering and tragedy of the troubles first-hand—and dedicate this book especially to them.

# Signs of War and Peace

# CHAPTER ONE

# History, Conflict, and Public Display in Northern Ireland

I first went to Northern Ireland to study Halloween, but I was struck (as I think most Americans are) by the preponderance of visual display. Brilliant multicolored murals adorn the walls and gable ends of houses; curbstones are in some neighborhoods painted red, white, and blue; in others, green, white, and gold. Starting in the spring, it seemed as if there was a parade almost every other day, and indeed, there well may have been; a Belfast city official told me there were 3,500 parades a year in Northern Ireland, which is about the size of Connecticut. The calendrical holidays in which I was interested, such as Halloween, had major public components such as bonfires and fireworks—and, again, parades. Likewise, the rites of passage of the life cycle have strongly visual public customs associated with them. Sometime before her wedding, for instance, a bride-to-be might be taken out to a party, gotten helplessly drunk, decorated with shaving cream, tin cans, and ribbons, and left tied to a tree or pole in some public area (Ballard 1998). Likewise, funerals customarily involve public processions of the coffin through the streets. Flags—British flags, Irish flags, Ulster flags—are ubiquitous. In addition to all this are the patrols of British soldiers, on foot and in armored vehicles through the streets, a visual and tangible presence of another sort.

I realized that a proper study of the conflict in Northern Ireland demanded close investigation of these and many other forms. It is immediately apparent upon arrival in Northern Ireland that there are "two sides" debating issues of national identity through these public displays. Whether the debate is limited to only two sides will be dealt with in some detail below, but that is certainly how the troubled political situation in Northern Ireland is characterized in the international media, and domestically as well. In Northern Ireland one frequently encounters the term "the two traditions," which refers broadly to British and Irish histories and heritages on the island of Ireland. The murals and the painted curbs encapsulate and encode these allegiances: Red, white, and blue colors reference the British Union flag and signify that the area is predominantly unionist, while the green, white, and gold refer to the flag of Eire, the Republic of Ireland; those areas displaying these colors are nationalist. The Irish Tricolour flag is actually green, white, and orange, but gold is frequently substituted. Likewise, while there are many different types of murals, most are self-evidently representative of one side or the other. Upon first impression, then, we encounter a split and divided society; the two sides each manifesting its existence, its identity, and its political positions in simplistic, if colorful, images.

## Significance

I lived with my family in Bangor, Northern Ireland (a seaside resort on the Belfast lough about 15 miles outside Belfast) from August 1991 to June 1992, after an initial visit in June 1990 sponsored by the British Council. My initial research project, as stated above, involved an ethnographic study of Halloween as practiced in Northern Ireland today (see Santino 1998). Because of all the activities mentioned above, and many others explored below, I expanded the scope of my work to include the uses of symbols in public more generally. I have returned every year since then (through 1996) for periods of up to six weeks, specifically to research these many forms of public rituals of presentation and display. In the course of doing so, it became obvious that apparently simple phenomena were complicated; that there are multiple distinctions among the spectatorship and among the generators and participants of these forms; that meanings are multiple, fluid, and shifting. I determined to investigate the ways in which public symbolic forms are used socially, how they have contextual and historical meanings and how these meanings are reconstructed and recreated in particular interactions and negotiations. I was interested in the parades and in the components of parades: banners, flags, musical instruments, costumes. I was interested in the cultural landscape: the murals, walls and curbstones; the fireworks displays and the effigy burnings; the spontaneous shrines. I thought I knew what I was looking for, and yet, due to my own ignorance and scholarly biases, I almost missed some of the most obvious and most displayed dimensions of culture and society.

For instance, during the summer of 1994, I watched as bonfire materials were collected and bonfires assembled throughout the province for the Eleventh of July. I attended the parades on the Twelfth, and on other dates. I interviewed participants and residents of varying ages and backgrounds. I attended historical reenactments. While conducting this research throughout the province, I noticed that several people were displaying red and black flags on their homes. Now, in certain areas, black flags commemorate the death of someone killed by a British soldier. Those flags are thus highly politicized and controversial. I was told, however, that these red and black flags were in support of a football team. I thought no more of them, even though I was interested in the use of flags as social display and continued to see them daily. A month or so later, however, back in the United States, I received a letter from Michael McCaughan, my friend and sometime collaborator who is a photographer and museum curator at the Ulster Folk and Transport Museum. He wrote to me concerning the victory of the County Down team in the all-Ireland Gaelic Athletic Association football final. He said: "The Down colours are red and black and coloured flags are widely displayed in the southern part of County Down. As you can imagine, there are none to be seen in Calvinist/non-GAA Newtownards. In short, the distribution of red and black flags reflects religious and ethnic identities in the county!"

McCaughan is precisely correct here. He pointed out to me a fact I had missed because I arbitrarily and artificially separated sports from politics, a major error. Team allegiances are frequently drawn along the Protestant-Catholic dichotomy. Moreover, by introducing the term Calvinist and referring to ethnic as well as religious identity, McCaughan problematizes the simplistic Protestant-Catholic terminology that is usually used to describe the situation in Northern Ireland. The "Calvinists" he refers to are Scots Presbyterians, and they, like people of English ancestry and the "native Irish," tend to adhere to different religions and live in more or less discrete regions.

There is more to the flag story, however. While the (Belfast) *Irish News* for September 19, 1994, headlined "Four in a Row" on the front page—followed by "Fans prepare to give champions all-star welcome"—on September 22 the *Newtownards Spectator* reported "Council snubs plea to congratulate GAA team." Refusing to congratulate a championship-winning local team seems unnecessarily boorish, even when one understands that the Gaelic Athletic League is universally considered a Catholic, Irish (as opposed to Protestant, British) organization. Indeed, even the sport—"Irish football"—is considered the cultural property, or "tradition," of the Catholic Irish. Still, in the interest of better relations (which almost everyone espouses), would not a congratulatory *beau geste* be at the least courteous, as well as politically prudent?

Again, the situation is more complicated than meets the eye. The newspaper report at first supports a perception of the refusal as a mark of ungracious intolerance:

There were noisy scenes at this week's council meeting in Bangor when Alliance [a political party that positions itself as nonsectarian and dedicated to amelioration of the schism] councillor Brian Wilson asked the council to send a letter of congratulations to the team.

His proposal was defeated by ten votes to four with the council's four Alliance representatives the only members to vote in favor.

Councillor Wilson said afterwards he was "disgusted" at what he described as "a very petty decision ... If we want to work towards peace then we have to respect the traditions of all communities."

This apparently very reasonable position is complicated, however, by the comments of another member of the council:

Independent Councillor Austin Lennon defended his stance in blocking the message of congratulations. He said the GAA's ban on security force members prevented him from congratulating the Down team.

"If the RUC [Royal Ulster Constabulary] are not good enough to serve with them," he added, "then I could not support them. I think they should have their doors open to all ... This is promoting a body that is not completely open."

At this point it appears that the council's position is not entirely unreasonable after all: The GAA itself apparently discriminates against some members of society. Then again, if one realizes that the RUC—that is, the police—has historically been Protestant in its make-up; that it has historically not protected the Catholic minority in Northern Ireland (in fact it has lent tacit, if not overt, support to violence perpetrated upon Catholics); that in fact it is perceived by Catholics, despite recent efforts to integrate it, as an enemy force, and not without reason, then the ban is understandable. This sequence of events and the rhetorical strategies that accompanied them, this social drama (see Turner 1974, 1982), indicates the complexities, the layers-of-the-onion quality of so much of life in Northern Ireland. Even the newspaper reports quoted above are full of coded significance for the readers. For instance, the use of the word "Irish" in the name *Irish News* is significant. It indicates an intended readership, a domain of coverage, and a generally Nationalist political position. It is not an accident that the victory of the (Northern) Irish County Down in the *All-Ireland* championship is celebrated in this paper. The victory of a Northern county is a statement that the six counties of Northern Ireland are in fact rightly a part of the republic. The Down team participates in the league as an Irish team, no different in that regard from teams in Cork, Limerick, and Galway. Meaning is signified and ethnic, religious, and political identities are expressed in a myriad of socially coded and understood ways: which newspaper one reads, one's nickname (for instance, Liam is a Catholic nickname for William; Will and Bill are Protestant); one's address and place of residence—indeed, even how one refers to Northern Ireland—all are significant.

These everyday signifiers are joined on occasion by the more overtly symbolic events and icons of the groups in question. For instance, the summer is called "the marching season" because it is then that the majority of parades occur, beginning on St. Patrick's Day for Catholics and Easter Monday for Protestants. The Protestant, or, more correctly, the unionist cycle reaches its high point on the Twelfth of July, the major festival of the summer for Protestants; it marks the Protestant victory of William III, Prince of Orange, over James II at the Battle of the Boyne in 1690, a victory that is said to have assured the British throne for Protestantism and led indirectly to the formation of a Protestant state in the north of Ireland. The week of July Twelfth and the following week are known as the Twelfth Fortnight, when many people are likely to go on summer vacation (or "holiday"). The parade season culminates with the procession of the Blackmen—an elite corps of the Orange Order, a fraternal Protestant organization that is fiercely unionist—in late August, and with the parade of the Ancient Order of Hibernians, a Catholic fraternal organization, on August 15 for the feast of the Assumption of the Blessed Virgin Mary. Tensions run very high during this period of time, and the routes of the Orange parades—indeed, the existence of the parades themselves—are challenged by Catholic residents who live in the neighborhoods through which the paraders walk, as on the Garvaghy Road in Portadown which became a flashpoint in the mid-1990s.

## Symbols in Everyday Life

In the summer of 1994 I was staying with a family in Bangor that is "mixed," in that the father is Catholic and the mother Protestant, but as Buckley and Kenney (1995:6) have noted, such mixed marriages are usually mixed in name only. That is, in such circumstances an agreement is reached in which one or the other religion is accepted as that of the entire family. In this case, Roman Catholicism prevailed. All the family's children attended Catholic school and Catholic church. It was in this context that I noted the following incident.

A boy who frequently played with the children of my host family came to the house on this day with a red, white, and blue baton. The children's father was incensed and sent him home. Later, he commented to me that he couldn't understand how the boy's parents could allow such a thing. The type of baton in question, of stylized union flag motifs, is widely sold prior to and during the period of the Twelfth. Youngsters can be seen at the parades carrying these toys, along with small flags and toy drums in emulation of the marching bands. The batons are very much an icon of the Twelfth of July celebration, which, as stated earlier, commemorates the victory of King William of Orange. William's victory is viewed by most Protestants as the decisive event that insured the British throne for Protestantism, saving it from the Roman Catholic threat and from the "error" of papism. The celebrations of the Twelfth of July, along with the bonfires the evening before, are widely perceived by people of varying backgrounds as celebrating the victory of Protestantism over Catholicism; in so doing, they ask that 45 percent of the Northern Irish population celebrate its own defeat. So when a child, playing with a toy that for him evoked associations of an enjoyable festival, went to play with his friends, presumably in innocence, he deeply offended his playmate's father, called into question his own upbringing, and was curtly dismissed from the premises.

These nuances that lie behind football flags and toy batons often escape the outsider. They are both historically based and socially generated; together they constitute a culture or cultures of stalemate, of contestation, of clash, and of desperate longing for peace. To get at them, we need an event-centered approach in which we look at both the cultural performances (such as parades and even situated forms such as murals) and the enactments that occur interpersonally; that is, the ways people actually use these symbolic forms vis-à-vis each other. The events covered in this book generally occur in public, but that is not to say that both private and personal considerations are not equally a part of the dynamics of the various constructions of meaning. We will see this most poignantly in the case of the spontaneous shrines that mark sudden, untimely deaths, often due to sectarian killings.

There has been a great deal written concerning contested space. In Northern Ireland the issues are not metaphorical, they are material. The land itself is Irish to some, British to others. Challenge and contestation over the right to define the nature of the place occur at all levels. Parade routes are met with organized resistance. Murals are whitewashed or defaced. People's interpretations of symbols are contradicted by counterinterpretations (see my discussion of Halloween bonfires, for instance, in Santino 1996). The metanarrative of history is constantly debated at all levels of society. The conflicts are internal to Ulster, but extend to include all of Ireland and Great Britain, and even to Canada and the United States. But in Ulster both the civilian population and the paramilitaries use the same forms of festive popular display—parades, murals, bonfires—to debate issues of power: Who controls the government, the economy; who gets to define the geopolitical state. Unionists and loyalists, though aligning themselves with the British government generally, are almost always thought of and referred to as "Irish" by the English. Along with this name come the stereotypes: Ulster unionists find themselves called "Paddy" and "Mick." They are identified with the group they most vigorously dissociate themselves from in Ulster, identified as such by those they most vigorously identify themselves with. Through their rituals of public display, unionists do more than state a public position. They desperately assert their British identity, only to see it rejected by those most British of all, the English.

Recent studies by Shane White, along with those of David Waldstreicher (1997) and Simon Newman (1997), of ritual, festival, celebration, and public display in the American colonies and during the early years of the republic, indicate a situation similar to that which exists in Northern Ireland today. These studies suggest that where there are contested nationalities, such expressive forms are particularly useful for asserting identity and claiming territory. These studies also help to redress an imbalance in the historical accounts that, prior to the works of LeRoy Ladurie (1979) and Natalie Zemon Davis (1975), tended to overlook folk and popular cultural events as lacking in historical (social) significance. Work originating in Northern Ireland itself (along with the Republic of Ireland, Great Britain, and other war-torn zones such as the Middle East) indicates that the political import of the public use of symbols is apparent to people who experience it on a daily basis. There is no need to justify its importance to them. But there is a responsibility to do more than document and describe these public activities.

The purpose of this book is to explicate some of the dynamics of public display performances in the context of guerilla war. The roles of these activities and symbolic forms are taken for granted in Northern Ireland but seem invisible to many of us in the United States. The guns and bombs, and the official negotiations, we witness on television and read about, briefly, in the newspapers. American politicians (who fully grasp the potency of theatrical political display in their own careers) make grandiose speeches about "lasting peace" and "justice" that are, in their generality, of little or no relevance to the residents of Northern Ireland. Politicians need to understand and recognize the role of culture as a concept that supersedes and envelops those of economics, politics, and the law, and I hope to make that case with this book. Each of these domains—the economy, the political and legal system—are themselves cultural artifacts, manifestations of specific cultural worldviews and ideologies. They are shaped by the ethos of the groups and practitioners who create them.

## A Troubled History

Edward D. Ives has written that the past is everything that ever happened. History, on the other hand, is a perceived coherence, "a backward look, the past as conceived and patterned in the present." History, then, "exists only in our minds ... only in the present ... [it] never was; it only is" (Ives 1988:5). The Irish have been accused of living in the past, of refusing to let go and get on with life. This formulation is intended to account for the ongoing troubles still afflicting Ireland, particularly in Ulster. However, it fails in at least two ways.

First, it fails to recognize that 55 percent of the population of Northern Ireland does not consider itself Irish, and as long as the world refuses to grant these people their own right of self-identification, the stalemate will continue. Secondly, it misses the essential presentistic nature of history as set forth by Ives. While both the Irish and the British residents of Northern Ireland have a very strong sense of historic grievance, they use the past to talk about the present. The historic battles that are constantly referred to are joined by more recent events, and all of these are used rhetorically and metaphorically to talk about, describe, explain, and situate people in an often frustrating, bewildering present that they did not create and in which they feel increasingly powerless. Thus, unionists in Derry say they are "still under siege," as a popular slogan has it, referencing the 1688–89 siege of Derry but also referring to current political situations.

The fact that history is contested is nowhere more evident than in Northern Ireland. Individuals point to different historic events as justification for present realities and present actions. Some historic events have diametrically opposed interpretation. My own rendering, then, despite my attempt at "objectivity," is also relativistic. Although I was born and raised Roman Catholic in Boston, Massachusetts, and my mother was of Irish descent, I do not feel a particular partisanship with the actions of the Irish Republican Army, and I think the British residents of Northern Ireland are the victims of an almost willful lack of understanding by the American press (when they receive any press at all). In the interests of disclosure, though, I must also say that my political sympathies are with the nationalists. If I were to support a political party in Northern Ireland, it would be the Social Democratic Labor Party, which, although fully nationalist, disavows the violence of the IRA (as opposed to Sinn Fein, which is associated with the IRA).

The "history" of Northern Ireland, then, is relative. It is, moreover, full of ironies, which the residents of Northern Ireland themselves frequently point out. For instance, during the Williamite wars, although King William of Orange has emerged as a George Washington-like figure for the Ulster Protestants, he—not the Catholic James—had the support of the pope. Probably the most significant event, in discussions of the troubles in Northern Ireland today, is the partition of Ireland in 1921; and this must be seen against almost a thousand years of imperial domination of Ireland by England. The Ireland of 1169 would hardly have been the Ireland we know today, of course, but instead an island of warring tribal states, when King Dermot invited English warrior soldiers to come in and help him in his battles with other tribes. At that time, the English pope, Adrian IV, granted

Henry II possession of Ireland, and by 1171 an English parliament had been established in Dublin.

Much later, in 1541, Henry VIII proclaimed himself king of the land of Ireland "as knit forever to the realm of England." Under Henry, the "Anglicization" of Ireland meant the large-scale removal of the native Irish from their lands in order to use them for sheep farming. Anglicization continued for a century, mostly in the north, where the land was confiscated and awarded to English and Scottish settlers. Although the northeastern-most corner of Ireland lies, on a clear day, within sight of Scotland across the Irish Sea, and people from Scotland had settled in Ireland for centuries, this "plantation" of Ireland was done with all the accoutrements of imperialism and racism. Land was taken from those whose families had lived and worked on it for generations. The native language was forbidden, as was the practicing of Roman Catholicism. The native Irish were visually depicted in English propaganda as subhuman, as apes; this strategic device allowed the English to maintain, to themselves at least, that their actions were justifiable as a civilizing project. Catholic uprisings were brutally suppressed by Oliver Cromwell, as at the massacres of Wexford and Drogheda in 1650.

Protestants were now in control of the state bureaucracy in Ireland, but in 1688 in London, King James was ousted from the throne due to his Catholic sympathies. He fled to Ireland, where he was pursued by the Dutch William of Orange who, it had been arranged, would replace James. Several battles that are today of great iconic significance in Northern Ireland took place during these campaigns on Irish soil, such as the siege of Derry. The forces of King James had surrounded the city of Derry and were starving the populace into surrender. The governor of Derry, Robert Lundy, decided to come to terms with James rather than see the people die; but as he went to meet with James, the apprentice boys of the city waylaid him and kept the gates of the city shut. Later, the forces of William arrived and routed those of James. Today, an effigy of Lundy is burned in Derry each December; the Apprentice Boys fraternal organization marches twice a year in commemoration of the siege; and the event has sparked some of the most potent slogans of the Northern unionists: "Still Under Siege" and "Never Surrender."

William defeated James at the Battle of the Boyne on July 12, 1690, and eventually solidified the Protestant, English-based ascendancy of power in Ireland. Eventually, under the Act of Union in 1800, power was centralized in London as the Irish parliament was dissolved. There followed many movements attempting to return to Irish Home Rule, which refers not to independence but to the presence of a British parliament in Ireland. In 1913, with the passage of a bill establishing Home Rule as imminent, northern Protestants who wished to preserve the union with Great Britain at all costs, and who feared that Home Rule would lead to a loss of their power in favor of the Catholic population, opposed Westminister. They formed the Ulster Volunteer Force and were prepared to fight the British—paradoxically, as Rolston has pointed out they were prepared to fight the British in the name of being British (1991:30). The formation of the UVF was met by the formation in the south of the Irish Volunteers.

During the first World War, Home Rule was suspended. The UVF joined the British army as the 36th Division, and was decimated at the Battle of the Somme. They are remembered in Ulster today as heroes and martyrs for their country. The war was problematic for Irish nationalists, for whom being part of Great Britain was anathema, but for many others the greater threat of the war was more compelling than the problems with England at home. The politics of separation were put aside temporarily and many, though not all, of the Irish Volunteers fought as members of the British army. Having put aside their political grievances and fought with Great Britain, many of these individuals felt betrayed by the British government upon their return to Ireland when no change in the status of the country was forthcoming. The more hard-line Irish Republican Brotherhood gained control of the Irish Volunteers, and seized control of the General Post Office in Dublin on Easter Monday, 1916. This rebellion against England was put down in one week, and did not, by all accounts, have widespread popular support until the leaders of the uprising were executed for their actions.

The uprising failed as such, but the strong anti-English sentiments that developed after the executions led to the Anglo-Irish War of 1919. A truce was declared in 1921, under which six northeastern counties of the province of Ulster would remain in the United Kingdom of Great Britain and Northern Ireland, while the rest of the country would become a British dominion with its own government. The deep division over whether to accept the partition led in turn to the Irish Civil War, but the treaty was signed in 1921; the Civil War continued until 1923. Although the forces favoring the treaty prevailed, the IRA remained intent on reunification.

In 1949, Ireland severed all remaining ties with Great Britain. Now known officially as Eire, southern Ireland claimed jurisdiction over the six counties of the North. The claim was institutionalized in its constitution, although it was unenforceable in reality. The laws of Northern Ireland, established as "a Protestant homeland for Protestant people"—much as Eire was "a Catholic homeland for Catholic people"—discriminated heavily against the Catholic population, who numbered roughly 40 percent of the total population of those six counties. Inspired in large part by the civil rights movement for African Americans in the United States, Catholics in the North initiated their own civil rights movement in 1968. Their political marches and demonstrations were met with wholesale attacks on Catholic neighborhoods, often with the implicit compliance of the police (such as when members of the RUC staged a general "sick-out" at a time when it was known that massive attacks on Catholics were being planned).

In fact, it was to protect the Catholic population that British soldiers were first deployed to Northern Ireland, where the Catholic residents cheered their arrival. After a year or so, however, it became obvious that the soldiers would use force against Catholic demonstrators. The soldiers quickly fell out of favor with that segment of the populace, and became targets of IRA attacks as well. Known as "provos," for "provisional," this IRA was a new generation of paramilitaries who became active as the elder members of the IRA of the 1920s passed away. It had no link to the government of Eire, and in fact was denounced by that government. Likewise, a newly reformed Ulster Volunteer Force, with no link to the original other than its name, arose in response to IRA activities.

The Irish "Troubles," then, as the conflict came to be known, can be viewed as having begun almost a millennium ago, or earlier in the twentieth century, or in 1969. Most date the modern Troubles from the violence of 1969; they have continued to the present day, despite the hopes offered by the cease-fires of 1995, which lasted for about a year.

Ireland is divided into four provinces. One of them, Ulster, is made up of nine counties. In these counties, much of the population is Protestant, many of them descendants of the plantation of Ulster in the seventeenth century (Robinson 1994). When Ireland gained dominion status in the British empire in 1921, six of Ulster's counties were partitioned and retained as part of Great Britain, forming the political entity known as Northern Ireland. These six counties have overall a majority Protestant population. The Republic of Ireland, or Eire, established after gaining full independence in 1949, claims these counties as part of its rightful and hereditary territory. As a result, there exists in Northern Ireland a kind of national identity crisis. Approximately 55 percent of the population are Protestants who are loyal to the British crown and who wish to retain the union with Great Britain. The remaining 45 percent are Roman Catholics, most of whom consider themselves Irish nationalists and would prefer a reunification with the Republic, or the South, as it is usually referred to. These divided loyalties, along with the ongoing presence of English troops, are the underlying context for the violence that is a part of everyday life in Northern Ireland. The contested control of territory on the international level (Ireland versus Great Britain) is echoed in the ritualization of space in Ulster.

The shootings, bombings, and killings are the most recognized aspects of the Troubles internationally. They are the most extreme manifestations of the deep social division, and they are despised by most residents of Northern Ireland. At bottom, the goals of the two primary interest groups—the nationalists and the unionists—are mutually exclusive, and so the more extreme manifestations of contestation continue as a kind of unfortunate, unpopular culture. The artistic displays support mutually contradictory claims. Spontaneous shrines are an important exception. In this sense the murals and parades are aspects of the conflict, regardless of whether they specifically address it.

This is not to suggest that they are subsidiary or secondary. The conflict in Ulster is dialogic and extends to all levels of society. I am not overlooking the fact that these various public display forms are found in great variety, each of which frames a kind of metacommunication, a way of framing thought about the conflict (it is heroic, it is tragic, it is defensive, it is biblical, and so on). These forms are multivocal and polysemic; different participants have different intentions at different times, and people derive various meanings, and kinds of meanings, from them in a number of ways. There is a multiple spectatorship for what amounts to competing visualities. In this book I am looking at the conflict holistically, placing the rituals, festivals, celebrations, and publicly displayed murals and shrines at the center of my consideration of the war in Northern Ireland.

## Forms of Public Display

The major forms we will examine include murals, parades, calendar celebrations, life cycle rituals, occasional celebrations (such as sports victories), and spontaneous shrines. These clearly overlap. Not only are they not mutually exclusive categories, but they are also not of the same level or order of analytical category. For instance, a parade can exist simply as a parade, or it can be one component of a larger celebratory or ritual event. Conversely, these forms can be broken down into other constituent components, also symbolically significant; examples are banners and musical instruments, which are used in parades, or flowers and personal memorabilia, used in shrines. So a drum might be a component of a parade, which in turn might be a component of a festival.

There is a great deal of material culture here: the shrines, with their flowers, mementos, and notes; the parades, with their banners, special clothing, and musical instruments; the painted walls and curbstones; the flags. The objects can generally stand alone, such as a flag flying on a pole, or be combined with other objects to create more complex units of meaning, a process I have termed assemblage. The materiality of these objects and events notwithstanding, in most cases we are dealing with kinesthetic processes: parading, beating a drum, wearing a uniform, and marching. We can look at the above listed categories from this perspective: Are they intended to be viewed as process or product? A parade is a process; a mural is a product.

All the events and artifacts listed above share this commonality: They are publicly displayed. With murals, however, the performance—the act of painting—is not the event to be viewed. We might see people painting or refurbishing a mural but it is not the primary focus. Murals form a background for the kinetic, processual, moving parades. However, murals are regularly painted over or retouched, so while they are immobile they are not exactly static. Parades, for their part, usually return to their starting points, so they are not exactly, or not entirely, linear. Both are related to cultural valorizations of time, and both tend to valorize space.

Calendar celebrations frame both parades and murals. They can be religious, such as saints' days; or historical commemorations, such as the anniversary of the Battle of the Boyne, Both types can be, and are, politicized. St. Patrick's Day is a nationalistic celebration for Catholics; the republican movement commemorates the Dublin Rising on Easter Monday. Protestants commemorate the Battle of the Boyne and the Siege of Derry in highly politicized and confrontational ways. In all these cases parades are central, and murals are painted and freshened. Also, as we have seen, sports allegiances are coded Protestant or Catholic, and so sports victory celebrations are also sites of intensification and exclusion, simultaneously.

Finally, the spontaneous shrines that mark the scene of an untimely death also represent the broadening of life-cycle ritual to a larger public audience, with ramifications beyond the usual ones to family and friends. Since they mark what are often sectarian killings, the deaths are of a heightened social and political nature. The public shrines address those ramifications and implications. Like murals, they are stationary—but also like murals, they are everchanging. Unlike murals, they are communal: Anyone who wishes to add a token may do so.

Throughout this book we will examine these materials according to a number of criteria. For instance, process versus product was mentioned above. And perhaps one of the keenest criteria employable here is the question of the commemorative versus the performative nature of the events.

The spontaneous shrines are the most clearly commemorative, because they are testaments to the dead. Still, though, they beseech the heavens and comment on earthly events. If by performative we simply mean acted, the performative aspects of the shrines are largely implicit, although such shrines often contain newspaper photos of the same shrines at earlier stages, reminding the public that the ritualistic actions that created these shrines were done by known people, and that these actions are contained, inscribed, in the shrines themselves. However, if by performative we mean, more correctly, actions that accomplish something socially (as J. L. Austin's early characterization had it for speech utterances that did what they said they were doing), then the shrines' performativity is to be found in the social arenas of paramilitary (and official) violence. In a discussion of the flower shrines for Princess Diana, Suzanne Greenhalgh (1999:51) points out that while the flowers commemorated a lost life, and expressed individuals' felt (if imagined) sense of relationship with the former princess, they also contested the style of royalty manifested by the Windsors and, in fact, altered it (at least temporarily). Thus, in that sense the Diana shrines were performative as well as commemorative. Likewise, the spontaneous shrines that commemorate violent death due to paramilitary action also comment on that violence, and the participants in the shrine hope to have an effect on the situation.

Murals, like shrines, are material artifacts. They are painted by unionists and loyalists, nationalists and republicans, for many purposes. Whereas the initial acts of painting are not ritualized, the murals themselves frequently are. They are painted for important calendrical celebrations, and retouched for these and other occasions. Likewise, curbstones are painted or repainted at these times. These painted surfaces are generally performative as well as celebratory, commemorative, and propagandistic. They serve as recruitment posters, as morale boosters, as territorial markers, and they have an active social role in the ongoing conflicts.

Banners, which are carried in parades for a number of occasions, are ceremonially initiated into use. After being created by a master craftsperson they "begin life" ritualistically, and their social contexts for use are almost always ritualistic. Like musical instruments, they are components of parades and demonstrations; but unlike instruments, they have a limited use value outside of ritual contexts.

The Lambeg drum is a curious case here. As an instrument, it too has its own "life." But it is an unusual instrument, heavily identified with the Protestant cause. When not carried and beaten in parades during the marching season of summer, the enormous, thundering Lambegs can be heard across the rolling countryside on summer nights, the echoes of drum competitions. These events are sanctioned by official agencies. Judges rate expertise, and the competitions are not ostensibly political. Still, to those "on the other side" who hear the drums, the sound is unwelcome, threatening, ominous. While there are exceptions—one example is the Catholic man, described later, who likes the sound of these drums—Catholics would not attend one of these competitions.

Parades are both commemorative and performative. Unionist parade participants maintain that their primary purposes are commemorative, celebratory, and traditional. However, the movement to contest parade routes, and the violent demonstrations and riots that have occurred along contested parade routes such as the Garvarghy road in Drumcree, Portadown, demon-strate the extent to which these parades are heavily significant, the ways in which they create social realities, the ways in which they are intrusive and martial.

The rites of passage have a public dimension to them—the bride gotten drunk, the coffin carried through the streets. These suggest, in addition to the performative-commemorative continuum, a public–private one as well. During rites of passage, the formal rituals are clearly performative. They are used to assign a social identity to a child, transform two individuals into a married couple, send someone off this mortal coil. The life cycle rituals that are most politicized are funerals. Paramilitary groups, particularly the IRA, perform military rites for fallen comrades in illegal but highly public funeral processions and burials. Moreover, Allen Feldman (1991) has suggested that much of the paramilitary experience—killing, being arrested, being interro-gated, and the entire Republican sequence of the blanket protest, the "dirty" protest, and the hunger strikes, were stylized and theatricalized rituals of passage within the paramilitary world. (See also Aretxaga, 1997, on women's protest experiences.)

The tangible, material aspects of the Ulster conflict are so prevalent that one prominent scholar has called his study *Material Conflicts* (Jarman

**Figure 1.   Irish Republican Army Funeral**

1997). But these objects are used by specific people in both particular and generally historicized contexts, and they are processual and dynamic. Objects can be used in contexts other than large-scale festivities, as when a flag is flown in order to irritate a neighbor of a different political persuasion. Throughout this book we will look at the ways in which symbols, as materialized in objects, painted on surfaces, and worn on the body, are combined with kinesthetic, patterned movement (such as parades, dances, or demonstrations) to effect communication, declare identity, stake positions, intimidate others, and fight an all-encompassing but not always violent war.

In *Formations of Violence,* Feldman says that "the exchange of violence is the principle economy of symbolic exchange in Northern Ireland" (1991:146). He demonstrates the theatricality of violence in Northern Ireland—and to some extent its ritualization. Examples include paramilitary actions, counterinsurgency activities, and interrogation procedures. I am examining the larger frames for all of this—the theatricality of life in Northern Ireland as experienced in parades, murals, and the rituals of death and politics (Santino 1992a).

Not only do these activities and artifacts communicate different messages and a variety of attitudes about the war; they are also all examples of the ritualization of space. Indeed, it is through ritualizing space that many profoundly important battles are fought and won, at least in the minds of the participants. Space—claiming territory and the power and right to name it, to traverse it, to celebrate it—is a recurrent trope in the North of Ireland.

Visual expressions of this schism are found throughout Northern Ireland. In working-class urban areas, people paint wall murals on the gable ends of buildings, and mark territory by painting curbstones the colors of either the British Union flag or the Irish Tricolour. Other examples include the display of the flags representative of one group. This often occurs in an area that is of mixed Catholic and Protestant population but where the group displaying the flags is in the majority, thus simultaneously signifying resistance and hegemony. Also, people use notes, photographs, and messages at spontaneous flower shrines constructed at the scenes of politically motivated killings to publicly express antiparamilitary sentiments in a way they might not otherwise feel safe doing.

Meaning—in symbols, in daily interaction, in public display—has been much studied. Symbolic anthropology has shown that symbols are multivocal and polysemic; that is, they speak in many "voices," such as color, texture, sound, and taste, and are perceived in differing ways according to the personal history of the observer and the contexts of the observation. Sociolinguistics has taught us to situate communicative events in social situations. Semiotics has directed us to deconstruct the ways in which symbolic forms signify meaning. Folklore studies have taught us that everything has a history and takes place in a context, and that what we think of as tradition is generative of socially specific new cultural forms. And cultural studies have forced us to come to grips with the politics of meaning formation, and shown us that meaning is contested: Who is in control of the symbolic form, and in whose best interests do specific readings serve? With the realization that meaning is socially constructed—that it is fluid, relative, and contested—how can we talk about meaning at all? How do we avoid privileging any one interpretation, especially the scholar's, but not only the scholar's?

For instance, the man who dismissed the boy over the baton, in the case discussed earlier in this chapter, was insulted by its presence on his property, while the boy's father felt that the baton was a harmless festive object, and that since no harm was intended, none should be inferred. The boy was oblivious to the adults' feelings; for him the baton was a pleasant icon of a summer festival. All the principal actors perceived and reacted to the same object but were situated differentially in relation to it. No one perception was "right," but all responses were genuine, all were "authentic"—the baton meant all those things, and provoked those behaviors, those social actions, accordingly. To complicate matters further, each of the individuals involved would say that the others' readings of the baton are objectively wrong, notwithstanding my own relativistic explication.

Simon Charsley has dealt with this issue in regard to wedding cakes (1987:93–110), choosing finally to render his own reading of a wedding cake as but one of many, no better and no worse than that of anyone else (including participants in the wedding) who might be differently informed, differently situated. I will argue that while all meanings are personal they are still socially constructed. The important word here is "socially," because, in the context of this discussion, the word implies the presence of history as a context of contemporary social meaning. Put simply: Not everyone will interpret a wedding cake the same way; few may see fertility symbolism in it, as the scholars usually do, but then again no one interprets it as a sign of Christmas, or Nazism, or bravery in battle, or witchcraft. In fact, many of Charsley's informants were in agreement with regard to certain aspects of the cake, such as its color—white—having something to do with purity, like the bride's dress.

So while symbols are multivocal and polysemic, interpretations are in large part determined not only by social situation but also by socioculture. Highly idiosyncratic interpretations are recognized as such. Interpretations vary; they may be oppositional and even contradictory, but they will usually be found to exist within a certain semiotic domain. A feminist reading of a wedding cake may see the color white as referencing purity—that is, virginity—and in turn as an offensive aspect of the patriarchal nature of wedding ceremonies. But the attitude taken toward the interpretation is not the same as the interpretation itself. The color of the cake, whether one approves, disapproves, or does not care, is still understood within a shared culture that is manifested in semantic and semiotic domains. Those domains are in large part historically determined, but historical narrative is itself fluid, open to challenge, question, and interpretation as well as rejection and replacement. History purports to uncover the past in order to help explain the present, but in fact it views distant events through the lens of recent ones. History will always be more about the present than the past. Nevertheless, the reason certain symbolic forms are felt to mean certain things and have social rules surrounding their usage, placement, and timing, is due to past events that are consciously understood and brought forth to justify the present.

## Terminology

The terminology can be confusing. The name "Ulster" is frequently used synonymously with Northern Ireland, but it is important to remember that the province of Ulster comprises nine counties, only six of which were included in the partition. The six counties are predominantly Protestant, but the three counties that are part of Ulster but not Northern Ireland include an Irish-speaking area, or *Gaeltacht,* in the heavily Catholic county of Donegal. Northern Ireland is internally divided: Protestant and Catholic, unionist and nationalist, loyalist and republican; and the historic province of Ulster is divided on a grander scale, being partially in the Republic of Ireland and partially in Northern Ireland.

Although the American perspective on Ireland usually casts the divisions as Protestant versus Catholic, this is simplistic. There are many Protestant denominations in Northern Ireland, principally the Church of Ireland (which is the Anglican Church in Ireland), Presbyterianism, Free Presbyterianism (a sect founded by the Reverend Ian Paisley), and a number of Fundamentalist movements as well. Likewise, in this context, it is specifically Roman Catholicism that is meant by those who denounce that religion for its purported idolatrous, anti-biblical liturgy and dogma. One might consider oneself a Catholic but mean that one is an Anglo- or Hiberno-Catholic; that is, a member of a "high church" (one with more emphasis on icon and ritual) such as the Church of England or the Church of Ireland. Thus it is more common to refer to someone as a Roman Catholic rather than simply a Catholic, as we are likely to do in the United States. It is the "Roman" part— the institution of the pope and what that represents—that is the issue.

The different denominations correspond to different ethnicities, different national origins. Presbyterians are Scots while Anglicans are descendants of the English. There are tensions between these two groups. Presbyterians have also experienced discrimination by the English, and at times they aligned themselves with the "native" Irish, the Roman Catholics, as for instance when members of both faiths joined the United Irishmen rebellion in 1798. However, by the twentieth century, all the various Protestant denominations shared a desire to remain in the union, and so they are now linked together as unionists and distinguished from Catholics according to this fundamental issue.

As in the case of Protestant versus Catholic, there are a series of parallel terminologies that are commonly brought into play when the Northern Ireland divisions are discussed. And like Protestant-Catholic, they are more complicated and nuanced than is usually understood. These can be juxtaposed to each other as follows:

| PROTESTANT | CATHOLIC |
|---|---|
| —Church of Ireland | Roman Catholic |
| —Presbyterians | |
| —Free Presbyterians | |
| —Fundamentalists | |
| Unionist | Nationalist |
| Loyalist | Republican |
| —Ulster Freedom Fighters (UFF) | —Irish Republican Army |
| —Ulster Volunteer Force (UVF) | —Irish Peoples' Liberation Organization (IPLO) |
| | —Irish National Liberation Association (INLA) |
| "Ulster" | "Irish" |
| Britain | Ireland |
| Teutonic | Celtic |
| Allegiance to Queen, union flag | Republic of Ireland, Tricolour flag |

The Protestant versus Catholic terminology has been discussed above. A unionist is a person who wishes to retain the union with Great Britain; a nationalist is someone who wishes to see the North reunited with the South. It is more accurate to describe the Ulster division in these political terms rather than the religious ones, although all these terminologies flow into each other. Indeed, Buckley and Kenney have argued that the sectarian religious groupings in Northern Ireland are essentially a kind of ethnicity; certainly religion functions as a source of identity politics in a way that race and ethnicity do in the United States and elsewhere (Buckley and Kenney 1995:14).

Among unionists, those who espouse the use of violence for what they perceive as self-defense, or as a means of achieving justice, or in direct retaliation to actions of the IRA, are known as loyalists; and those nationalists who espouse violent (or paramilitary) actions are called republicans. An attack on citizens that is said to be sectarian in nature is generally motivated by the political beliefs of the parties involved, although randomly retaliatory or "tit for tat" killings might target members of the "opposite" religious faith. Still, it is more accurate to say that most attacks involve loyalists or republicans rather than Protestants or Catholics. Loyalists have spawned several paramilitary groups; the Ulster Volunteer Force (UVF), the Ulster Defense Association (UDA), and the Ulster Freedom Fighters (UFF) are the most prominent. The Irish Republican Army (IRA) is the primary republican paramilitary organization, although it has even more militant splinter groups within it, such as the Irish Peoples' Liberation Organization (it is no accident that the initials also reference the Palestine Liberation Organization) and the Irish National Liberation Association. After the Good Friday peace agreement, a dissident group calling itself the Real IRA exploded the Omagh bomb which killed 26 people and caused some of the worst devastation seen in Ireland.

Unionists and loyalists tend to use the term "Ulster" when naming organizations and otherwise referring to their formal activities, whereas nationalists and republicans use the term "Irish." Thus we have the Ulster Volunteer Force and the Irish Republican Army. Unionists define themselves as British; they swear allegiance to the queen and the British flag (known popularly as the Union Jack, officially as the union flag). Nationalists consider themselves Irish, and look to the Republic of Ireland and its Tricolour flag as their symbols of nationhood and identity, the emblems of their imagined community (Anderson 1983). Unionists see themselves as a branch of the Teutonic peoples, and will categorize their cultural traditions (such as parades) as Teutonic; nationalists define themselves as Celtic. Despite the fact that nationalists, Catholics, and native Irish peoples also march in parades and wear collarettes much like the unionists (the colors are different but the organizational paradigm is the same), the traditions most typically articulated as "Celtic" and as "belonging" to this side are instrumental music and song (what we and they think of as "Irish traditional music") along with storytelling and the verbal art known generally as "crack."

While these terms and my definitions of them are imperfect—they are used somewhat ambiguously in Ireland itself, and membership overlaps among the categories—they are indicators of the cultural and cognitive taxonomies in Northern Ireland, and also the Republic of Ireland and Great Britain. The distinctions are important: A nationalist is not necessarily a supporter of the IRA, for instance, and in fact the Social Democrat and Labor Party and Sinn Fein are political rivals, competing for votes among the nationalist citizenry precisely over this issue. We will see, however, that a great deal of the war—and that is what it is—in Northern Ireland is not fought with bullets or bombs but with (or through) the public display of symbolic forms. Also, we will see that these terrains are themselves contested, and that the symbols belonging to one side are sometimes claimed by the other. On one level, for instance, both the Catholic Church and the Church of Ireland claim St. Patrick as their patron saint, while in mural paintings loyalists claim the figure of Cuchullain (an epic hero long considered a Celtic warrior and thus the cultural "property" of nationalism) as a pre-Celtic resident of Ulster who repelled the invading Celts. We will consider this later at greater length, but the point is that not only are political points of view established through murals, or territory claimed through parades, but that the nature of the symbolic forms and activities are themselves challenged and contested. The use of the same image by otherwise combative groups is deceptive. An outsider might view it as a point of commonality, when in fact it is particularly contestive.

Many of the categories listed above can be viewed as transformations of the others. Catholics are generally nationalists, for instance, and Protestants are generally unionists. In some ways however, to substitute one category for another in the scheme outlined above is a serious mistake, one that outsiders frequently make. That is, the problems are frequently portrayed in the international media as simultaneously Protestant versus Catholic and English versus Irish. In this blurring of categories, Protestantism becomes a synecdoche for England, Catholicism for Ireland, thus conflating—and confusing—religion, ethnicity, and political affiliation.

Sometimes this transformation has an intermediary stage, such as Northern Ireland versus the Republic of Ireland; here Northern Ireland stands for England and Protestantism (and by extension Great Britain and the Teutonic peoples), while Catholicism becomes Ireland becomes the Celtic peoples (which, ironically, includes the Scottish). These suppositions obscure the real complexities of the place and lead to stereotyping. Further, it is a mistake of order, substituting a category or classification that exists on one social level with that of another level. It is a kind of magical thinking—to effect a part is to effect the whole. Too often, I believe, political attempts at resolving the problems in Northern Ireland have failed because politicians have operated on this principle of substitution, on the erroneous assumption that to effect agreement on one level is to effect change on all levels. Dublin and London may sign the Anglo-Irish Agreement, for instance, but this does not mean that interpersonal and social relationships in the streets and neighborhoods of Belfast improve, much as we might wish otherwise.

## Celebrations and Politics

It is a mistake to view an Orange parade, an IRA funeral, or a Bloody Sunday commemoration as second-order cultural phenomena, as a reflection of the culture. Again, the population of Northern Ireland knows better. Because of their real potency and significance, these events and others like them are fiercely contested. Witness the "Reroute Sectarian Parades" movement, for instance. Orange parade routes through nationalist or republican areas are so controversial that the resultant civil disturbances surrounding them, as at Drumcree in the late 1990s, threaten to permanently derail peace settlement negotiations. Nevertheless, these events are always much more than political statements. Even the explicitly political rallies are also festive, even carnivalesque events (see for instance Buckley and Kenney 1995:153–172). They are simultaneously rites of intensification of ethnic identity, the construction and maintenance of which always involve the construction of differential identity, the creating of an "other" against which to define oneself.

For instance, during the hated period when Catholicism was outlawed, priests and lay people risked their lives to hold masses at large rocks, known as "mass rocks." These mass rocks are still found throughout Ireland and are emblematic of that oppression. For the past 20 years one such rock has been revived as a site for the sacred mass. Interestingly, this ritual, reeking with historical and political reference, is conducted on the first Sunday in August because, according to one member of the Ancient Order of Hibernians, "It's always been sort of a set day for us." In fact, August First is one of the four quarter days in pre-Christian Irish society known as Lughnasa, so the use of this date not only refers to the suffering endured during Protestant British rule, but also reminds people of the "native" status of the Catholic Irish dating back to Celtic times. Tradition is used here to generate a celebration that is simultaneously deeply contextualized by history and very much a contemporary social statement.

Other events are also very much part of current social dialogue. Some of these events are holidays, such as the Twelfth of July. Some are religious, historical, or ethnic commemorations, such as St. Patrick's Day parades or the Apprentice Boys' parade to commemorate the Siege of Derry. Some are life-cycle rites, such as funerals. These events, along with political demonstrations, are discrete events with their own overt purposes. They should not be thought of as all of a kind. Neither, however, are they single-focused in nature. For instance, while researching the bonfires of the Eleventh of July, I had visited several bonfire sites during the period of several weeks of their construction, and then on the night of their conflagration. The combustible materials are gathered for weeks in advance of the Eleventh night, as it is called, by adolescent males. These boys frequently guard their bonfire materials against potential raids from other groups of boys who hope to thwart their attempt to build the biggest, most impressive bonfire. Huts are built adjacent to the actual bonfire structure, or space is made inside the structure itself for boys to sleep in overnight as a precaution against rivals. The bonfires are built on any available space, which frequently means that the large fires are uncomfortably close to people's residences. More likely, though, they are built in a large open field, and if space exists on a high place in the topography, so much the better. On the Eleventh night itself, one can see fires dotting the landscape for long distances from atop a hill. I was struck by the total engagement these boys and young men—ranging in age from six the to mid-twenties—had in their bonfire projects, the way they evaluated them aesthetically and took pride in them, the way they used surplus materials that the community had no real use for, and how those materials were thus disposed of in a way that was constitutive of community. There is much about these bonfires that is of positive community value.

I even thought about how ancient a tradition summer bonfires were in both Ireland and throughout the British Isles. The pre-Christian Celtic ritual repertoire certainly included bonfires as a major component, as did the British (Cressy 1989; Santino 1996). However, on these contemporary bonfires were effigies of the pope, flags of the Republic of Ireland, and greenery symbolizing the Irish, to be destroyed in the flames. This sectarianism is very much a part of the Eleventh night fires. More than the parades the next

**Figure 2.  Eleventh Night Bonfire**

day, the bonfires are thought to be dangerous events that should be avoided by Roman Catholics. The problem, then, is that the bonfire tradition, while composed of many positive aspects, is also politicized in an ugly way. These two facets seem to be hopelessly entwined, but they were not always so and perhaps may not be in the future. For now, though, we see what Victor Turner (1967:28–30) has described as a conflation of the ideational with the sensory: That is, the ritual actions are sensually pleasant—it is fun to gather the combustible materials, to camp out at the site, to enjoy the fire on a summer night. These positive feelings are linked to anti-Catholicism and reinforce the widespread sense of belonging to a culture that is "under siege" and a heritage that requires defending. Some people suggest doing away with the bonfires legislatively, as if this would erase the sectarianism they are used to express. But as I have tried to demonstrate, the appeal of the large outdoor fires is as broad as it is deep. They are not only about sectarianism and political identity, though these dimensions are very much there. They are also about bonding with friends, about self-esteem, about family and neighborhood and even season. To legislate against the bonfires would cause deep resentment; to do away with them, even if possible, would be to do away with all of this.

For a scholar interested in ritual-display issues that are of direct social and political impact, no place can serve as a better site for research than Northern Ireland. But there is much more to it than scholarly interest, as important as that is. Almost all of the people in Northern Ireland want peace more than any other thing. The rituals, festivals, celebrations, and public displays that I discuss in this book are all artistic genres, and we will see that in Northern Ireland, art and politics are one. Playing traditional Irish music, dancing a jig, entering a drum competition, are all, whether intentionally or not, political acts in the North. The popular and folk expressive forms are seen as manifestations of social and political identity— and also, as we shall see, an imagined racial identity. Moreover, life is seen by many as a zero sum game. If one group publicly displays its popular art, it is seen as a defeat for the other. In Northern Ireland, life and art (at the least, popular and folk art) and research are all enormously compelling, because the issues, events, and people are at once emotionally stirring and intellectually exciting: The ritual display in Northern Ireland is artistic, expressive, impressive, and urgent. It is about war and death, peace and life.

# CHAPTER TWO

# Ritual Display and Presentation

In Northern Ireland there are many contexts in which to situate public display, and many histories to examine. On June 2, 1995, I drove with Michael McCaughan to Dungiven, County Derry. We were visiting a rag well, a holy well with healing properties. This one was surrounded, indeed almost obscured, by trees and bushes on which were tied rags, strings, ropes, and other pieces of cloth left by previous visitors as votive offerings. The well is located close to the ruins of the medieval priory of St. Mary, overlooking a spectacular view of the valley below. The well is easily missed as one walks onto the grounds, but once seen, it is unforgettable. The branches are thick, twisted, gnarled, and dense, made all the denser by the great many rags tied on them. It is both a wall made of rags and branches, and an environment, a space into which one must enter to get to the small rock basin with its holy water. Most of the rags are faded and in various stages of disintegration. Perhaps people leave these in the belief that as the rags disintegrate, the illness to which they correspond will also fade. However, Michael and I found the site both striking and moving, and we each left a token of our presence—I tore my handkerchief in two pieces that we tied to branches of the hawthornes, which are said to have supernatural associations of their own. Broadly speaking, the rags are at least in part, memorials—tokens of our having visited. Clearly the motivations that lead to such public acts of memorialization are many and complex, having to do with sickness, belief, personal devotion, attempts to influence that which is beyond human control, and also a need to demonstrate to an unknown audience that one participated, that one contributed to this monument, that one was here, albeit anonymously. Both the commemorative and the performative are seen in this holy rag well.

Nearby the ancient priory lies in ruins, but inside its walls there is a medieval stone tomb with a recumbant sculpted figure lying atop it. A monument to an ancient death, memorialized, lost in time but still with us, still signifying. The rags will disintegrate far more quickly than the stone will crumble, and this permanence or lack of it is more than simply a byproduct of the materials. It is a factor in the aesthetic quality of the memorials themselves. Both the priory and the rag well are profound as places where people have confronted and still confront issues of death and suffering, and leave material traces, public tokens, as evidence of belief, actions, and private emotions.

Ireland is justifiably famous for its musicians, its poets, and its story-tellers. However, despite the fact that Ireland is internationally revered for its oral and musical traditions, there are many visual aspects to Irish culture. The beautiful Celtic designs in illuminated manuscripts such as the Book of Kells, for instance, or found decoratively on the Iron Age shields and weapons of ancient warriors, speak to this visual aesthetic in what we now call Celtic culture. More relevant to this study are the many forms of visual public display associated with ritual, festival, and celebration that have been reported in Ireland and throughout the supposed Celtic cultural area. While recognizing that the concept of a unified Celtic culture is a romantic fiction, we can still note that in Galicia—an area of Spain that is said to have a Celtic heritage and that was the site of the medieval pilgrimage to the shrine of Santiago de Compostela (St. James)—pilgrims left small piles of stones known as milladoiros along the way, marking their passage (see Frey 1998:23–26). (Recently a Celtic revival group of Galician musicians have taken this term as the name of their band.) Today similar practices are associated with death and internment rituals in Ireland.

Rituals of the life cycle have had and still have important public components in Ireland. As described earlier the display of pieces of cloth left at a holy well by people who have made a pilgrimage there is one example. Carrying the coffin of a deceased person from the home to the church is another. Likewise, the calendrical festivals of the yearly cycle are marked with mumming processions such as those of the Wren Boys in the winter or the St. Bridget processions on the feast of that saint in February. People celebrate other occasions such as St. Patrick's Day by attending parades, and many people on that day attach a sprig of shamrocks to their clothing. Moreover, on a day of intense Irish nationalism such as St. Patrick's Day, people frequently dress in green while businesses and domestic buildings are decorated with bunting and flags.

Public decorating for and ritual marking of special times and places is of course a well-known phenomenon internationally. Beyond the obvious examples of domestic and institutional decoration for important religious and calendrical occasions should be added the tendency to ritualize and mark absent presences—for example, using yellow ribbons to denote hostages or green willows to denote absent lovers. Likewise, the deceased have been ritualized in most societies for which we have evidence. When death is sudden, untimely, or unexpected, as in the case of automobile accidents, for instance, some cultures have found it appropriate to mark the place where the death occurred (see Gillis 1994 for a general discussion of commemorations). Roadside crosses are found throughout the American Southwest and Latin America, for instance, as well as in many European countries. In the last decades of the twentieth century the custom of marking the place of a shocking death with a spontaneous shrine consisting of flowers and personal memorabilia has grown from a regional tradition to become part of the global expressive repertoire. The entrance to the Dakota Hotel in New York became such a memory site after the assassination of John Lennon, as did the apartment building of Gianni Versace in Miami Beach after he suffered a similar fate. In Spain, the paramilitary Basque Separatist group ETA kidnapped a young part-time rural politician, Miguel Angel Blanco, demanding that political prisoners be interned in prisons closer to their homes. The group threatened to kill their hostage if their demands were not met. There was widespread revulsion at this act, as expressed through the display of blue ribbons. In July 1997, the ETA did in fact execute their prisoner, and the blue ribbons were replaced with black. Moreover, spontaneous flower shrines to the memory of Blanco were erected in Madrid and throughout Spain, and marches, rallies, and demonstrations protesting the ETA were joined even by other groups who favored their cause but not their actions.

Examples are legion in the contemporary world. From the Mourning Wall in Oklahoma City in response to the bombing of a federal building, to memorial walls in other cities commemorating gang member deaths, to the "floral revolution" in London and Paris in response to the death of Princess Diana (see Walter 1999, and Kear and Sternberg 1999), spontaneous shrines and the public marking of the places where death has occurred have become a primary response. Concomitantly, newspapers and other media frequently (ritualistically?) feature these shrines as part of their coverage of these events. How this tradition has coalesced and spread at this point in history is an important question (see Santino 1992b), but we should remember that the actual instances of ritual commemoration are specific to particular times, places, cultures, societies, and people. In Northern Ireland, while fatal roadside crashes are indeed noted and sanctified with crosses, flowers, and wreaths, by far the majority of such shrines are associated with killings motivated by political conflict and hatred. Spontaneous shrines in Northern Ireland are thus, for the most part, products of and speak to the Troubles, the guerrilla war that divides that society. We will return to these shrines in some detail in a later chapter.

All these events, the intentions of the actors, the meanings they and others ascribe to the actions and to the things they leave behind, are not the same; I am not equating them. Rather I am suggesting that ancient and traditional paradigms of public display are very much a part of traditional life in Ireland, and that while these paradigms have a broad relevance internationally (perhaps due to the media-dominated nature of contemporary culture and the visual aspects of the displays), they have developed particular significance in Northern Ireland.

## Public Display

We can and should separate the concept of "public" from that of "display." Clearly the concept of public display is a broad one. Roger Abrahams suggested it with reference to events such as "the parade, the pageant, or the ball game … expositions and meets, games and carnivals and auctions" (1981:303–304). Abrahams uses the term to refer to "planned-for public occasions" in which "accumulated feelings may be channeled into contest, drama, or some other form of display," and which also includes those "actions and objects [which] are invested with meaning and values [and] are put on display" (1981:303). In a similar but not identical way, Don Handelman has suggested that we deal with the problem of imprecise terminologies such as ritual, festival, spectacle, rite, and so forth, by referring to these phenomena as public events, and then developing a taxonomy and analytical methodology for each event based on its design; that is, what the particular event is intended to accomplish socially (Handelman 1990). In a study of a series of "living celebrations" that I curated for the Smithsonian Institution in the early 1980s, I used the term "rites of public presentation" to describe the staged presentations of traditional cultures; these were designed in part by the participants, who were members of the community whose traditions were being represented. I made the point that while the nature of each event was necessarily transformed, they were all ritualistic in a different way. Not simply because they were theatrical, but because while they were presented to the public as a kind of entertainment (however edifying), in practice they turned out to be something else, a hybrid form during which audiences actually participated in rites that were foreign to them. The role of the audience members was transformed from that of passive viewers of a theatricalized event to active ritual participants (Santino 1988).

Samuel Kinser cites the slaughtering of "steers and other animals before the Pope and other Roman notables after a parade through the city," as reported in a document circa 1140 A.D., as the earliest report of festive customs prior to Lent, that is, Mardi Gras or carnival (Kinser 1990:3). This is clearly an expressive public event in that it is a festive or ritual act carried out in a public place. Public places can be either indoors or outdoors—a church or a plaza, for instance. The concept of "public," as I am using it, generally implies unrestricted public access to an event that is often but not necessarily held outdoors. In many rituals, audiences can be restricted to, say, men only, or to friends and relatives of the primary participants. In a sense all rituals are display events, in that an audience of some kind is necessary to witness and validate the changes wrought by the ritual or the claims it stakes; but not all rituals are necessarily public. The term "public" here has to do with spectatorship, rather than a Habermassian sense of shared civic interest.

Here we think of ritual as social dramatic enactments that the participants believe to have some transformational or confirmatory agency and to derive this power from an overarching parahuman authority such as a deity, the state, an institution such as a university, and so on; rather than ritual in the sense of custom, or even more broadly, routine. That is, rituals, while not always religious, are always sacred (Moore and Myerhoff 1977). The authenticating authority in religious rituals is thought to be divine, but in secular rituals—for instance, a commencement exercise—it is the scholarly institution, in the person of its representative, that validates the transaction. It is understood that an individual—a university president, say—is ultimately only a transitory representative of a conceptual but no less real social institutional entity. An individual in such a role is a sign of the institution, and the institution transcends any one individual. Thus, in commencements, as in inaugurations, retirements, and so on, while the individuals involved do not necessarily consider these to be religious ceremonies, they do recognize an authority greater than themselves and a matrix derived along political, economic, or institutional domains that is analogous to the position of religion in specifically religious rituals—while not necessarily like religion in any other way; I am not making the case that these are "secular religions."

In such rituals—ceremonies recognized as such by both the participants involved and the society generally—there is a component of public recognition and witnessing of the transformation that occurs. This is most obvious in rituals of the life cycle, since such transitions directly affect the entire social network of the individuals involved and the community of which they are a part. So friends and relatives are present at Bar and Bat Mitzvahs, weddings, funerals, *quinceañeras,* and retirement parties, confirming social recognition on the change of status of the individuals involved while acknowledging their own concomitant change in social status vis-à-vis the individuals involved. But at the same time, these events are restricted to an invited guest list. Display is an important dimension in these events, but to what extent are they public?

The rag well described earlier is a prime example of a custom that involves public display, but the well itself is hidden and hard to find. One must seek it out. Thus the clientele for this healing well is self-selected. On the other hand, a parade in a city or town will be witnessed by many people who have not deliberately set out to watch it. Similarly, a display of festive decorations on the façade of a building will be seen by an indiscriminate audience of passersby. The events and artifacts examined throughout this book such as parades, murals, and spontaneous shrines are generally witnessed by an unspecified and unrestricted—indeed, unknowable—audience. The concept of a public-display event is admittedly broad; but by treating the ritual genres as display events or as rituals of public presentation, we can take a performance-centered approach to examining the symbolic associations, the contingencies and contexts of presentation along with larger frames of time, place, and proximity, and also those of race, gender, politics, and power. Another important axis for examining such events, then, is that of restricted versus unrestricted viewership or participation. The issue of the public to whom an event is displayed is one that needs to be problematized.

## A Geography of Public Display in Northern Ireland

In the past, comprehensive accounts of traditional public ritual display events tended to be temporal in their organization. Kevin Danaher's *The Year in Ireland* (1972) is an excellent example. Likewise, R. H. Buchanan's "Calendar Customs" (Buchanan 1962; 1963) is another. I have written a similarly organized volume concerning American calendar customs (Santino 1994a); in addition, my study of Halloween in Ulster (1998) situates that festival in the context of other customs and celebrations in seasonal proximity, such as harvest celebrations and Guy Fawkes Night. In this book, we are examining public space, including festive events such as parades and fireworks displays, and also the marked culturescape (McCaughan, personal communication; see also Appadurai 1990). By "marked culturescape" I am referring not only to the ways people have shaped the land, but also to the many markings left on the environment, both built and natural. I mean, quite simply, the painted walls, curbstones, graffiti, and murals that are ubiquitous in the towns, cities, and even the countryside of Ulster. These have to do with geography in the very direct sense of territoriality.

## The Lambeg Drum

The territoriality of the parades is such that Orange parades and Green parades, according to Neil Jarman, never traverse the same roads (1997:139). Further, the Orange parades had exclusive access to center city Belfast and Derry. It was not until the Sinn Fein demonstration of August 1994—also a parade, or perhaps more accurately a procession, political in nature—that

**Figure 3. Lambeg Drum with Orange Lilies Attached**

the Republicans triumphantly marched into the middle of Belfast, to City Hall, where Sinn Fein leader Gerry Adams placed an Irish flag on the building and declared it liberated. These actions were entirely symbolic in that City Hall went back to business as usual the next day—this was not a conquering army. Such actions, however, show the importance of street theater in these circumstances. In an analogous way, the Reverend Ian Paisley, an outspoken, hard-line unionist, has been known to bring a Lambeg drum to a demonstration at City Hall. Lambeg drums produce a deafening roar. They are said (perhaps erroneously) to have been introduced into Ulster by King William's soldiers, and they are frequently adorned with flowers such as sweetwilliam and orange lilies, both of which represent William of Orange. Like unwelcome parades, the deafening noise of this drum penetrates private space; there is no escaping it. The beating of the drum essentially amounts to an aural taunt, a kind of rough music that is to Catholics the equivalent of, or worse than, the anti-Catholic chants sometimes heard at other events. On August 16, 1992, I asked Reverend Paisley why he carried the Lambeg drum with him during his protest. The following dialogue ensued:

I.P.: Well, the Lambeg drum is what we call a "lasher," is the colloquial term for it. And it has always been, it has always been a symbol of rallying.

Tradition says it was a Lambeg drum that rallied the troops before the [Battle of the] Boyne. And the thunder of it, the, the liveliness of it would probably bear that out. It could be heard over a long distance. It was like a room full of thunder, more or less. So, in demonstrations, most demonstrations in the olden days there was always Lambeg drums. I think the fact that there's not so many of them today is because processions move at a rate that it's impossible for a Lambeg drummer to keep up with. They're a pretty weighty instrument and so they are better just for more or less a demonstration rather than a parade, although some of them are still in the Orange parades.

J.S.: Did you carry it yourself?

I.P.: Well, I did beat that drum, yes, and I beat the Lambeg drum on many occasions ... I think that my—I think there was a painting of myself on that drum. It was a usual thing that leaders had their portraits painted on drums.

J.S.: Would you have known the individual who painted the portrait?

I.P.: Oh yes.

J.S.: Would the drum have been presented to you? Was it your drum?

I.P.: No, no. It was his drum, but my painting on it.

J.S.: So what kind of statement, what are you thinking when you say, "I'm going to bring the drum to this"?

I.P.: Well, I think it's just an identification with the historical connection of what this whole struggle is about and also an identifying of the people with the struggle.

The Reverend Paisley chooses to lead a demonstration at city hall bearing the enormous Lambeg drum, a drum that he associates with military battle and victory, with the defeat of the native Catholic Irish at the Battle of the Boyne. Certainly his nationalist opponents along with everyday Roman Catholic citizens could understandably read offensive, threatening, and gloating (what the Irish refer to as "coattrailing") symbolism in this act. While Reverend Paisley denies these intentions, one can see the ways this act and so many others appear to be ritualized battles themselves; they are dramatic enactments of victory and defeat, symbolic microcosms of the war that continues around the participants. In regard to this act, however, the drum is also ceremonially carried by the person whose portrait adorns it. To Paisley, this fact must have been equally as important as the act itself, and suffused the drum with a different kind of meaning for him. These symbolic gestures and objects—procession, flags, drum—have great social power, and are not taken lightly.

In addition to murals that are retouched for the Twelfth of July, the curbstones are painted in the colors of one or the other of the two national flags—Great Britain's and Ireland's—that are in contestation. Usually painted by young boys in advance of important celebrations, particularly the Twelfth of July, the curbstones mark neighborhoods and territory throughout the year. Like many of the activities we will look at in this book,

Figure 4.  Mural with Painted Curbstone

this is largely innocent. Participating in the customary activities associated with the Twelfth and other calendrical festivals is fun. Watching parades, marching in parades, gathering combustibles and igniting them in bonfires—stripped of political and sectarian dimensions, these activities are in and of themselves engaging and enjoyable for young people. In advance of the Twelfth, residents of city neighborhoods spruce up their blocks in competition with each other to see who has the best decorations. It is in this context that, along with putting up bunting and flags, the sidewalk curbs are painted anew.

Not all residents approve. One woman wrote to the *Belfast Telegraph* in July 1994, angrily complaining of the practice of curbstone painting. Her reasoning was that the colors used represent the union flag, the flag of Northern Ireland, and as such do not belong in the gutter. From a different perspective, Mr. John Joe Bradley, who has been a lifelong member of the Ancient Order of Hibernians but who insists that he takes no offense at most of the unionist public displays, including parades, bonfires, and the Lambeg drums, objects vigorously to the painted curbs:

J.J.B.: I think it's terrible, painting your curbstones. I think it's uncalled for.
J.S.: Why?

J.J.B.: Because it maps out an area. That we [the Orange Order, or unionists] are here, and we're dominant here, and as far as I'm concerned, none of us are dominant anywhere. We're all on this earth for a certain amount of time and it shouldn't be there. I don't think that's important and I don't think it should be done.

J.S.: And the arch, you don't think is a problem?

J.J.B.: The arch, the arch you see was there for maybe two hundred years. It's part of the North for us. The curbstone painting is a new thing. We probably haven't had the curbstones for two hundred years, that's a new thing that came out. The bunting is up. But if we [the AOH, or nationalists] put up bunting—as a matter of fact, we have a major parade every single year on the Fifteenth of August [for the Feast of the Assumption of the Blessed Virgin Mary]. I would put up the bunting on the fourteenth evening and I would take it down on the sixteenth morning. But I think the unionists and the Orange [Order] would let it stay up for a couple of weeks, three weeks, for whatever reason. I don't know, that's their business. It doesn't give me offense.

There is an air of festival in the streets as the Twelfth approaches. Nevertheless, there is at the same time a pronounced tension between Protestants and Catholics as the height of the marching season nears. Many people report that long-time friends, neighbors, and workmates stop talking to each other. The Twelfth of July is simultaneously carnival for many and offensive display for others. Participants usually feign ignorance of any offense and claim not to understand it, emphasizing the festive aspects and their pleasure in it. Some Catholics speak of having enjoyed the parades in their youth when they were unaware of their political aspects. On the other hand, Protestants typically speak of the old days, before the Catholic civil rights disturbances of the late 1960s and early 1970s, when Catholic neighbors would tend to the chores for their Protestant neighbors so that they could participate and enjoy the Twelfth; or when Catholics themselves would enjoy the parades. But as many Catholic people told me, they feel that this has been exaggerated, and does not reflect reality so much as wishful thinking on the part of the Protestant population.

For instance, I interviewed a 35-year-old Catholic woman named Maureen who had become a friend of ours. As a child, Maureen looked forward to the Twelfth of July because of all the bands. Although she went to a different school from her Protestant friends, she mixed freely with them and some of them were in the bands. It was fun for her to see them. Today as they march down the Brunswick Road they make a special racket when they pass by the Catholic church. "It takes more out of them than it does out of me," she says.

Several times during our conversation she repeated that society was not as divided in her childhood as it is today. Maureen was born just prior to the most recent phase of the Troubles. She says, "When I got older I found out what it [the Twelfth of July] was." She repeated this several times during the conversation, but said that finding out what it was did not make her feel any more awkward about attending the parades. On the other hand, she also told me that up until recent times there were Eleventh night bonfires in Bangor by the castle, in the center of town. She never attended these, and her body language and facial expression indicated disdain as she spoke of them. Similarly she said that the fireworks used at this time have gotten increasingly dangerous, implying that this is more evidence of the increasingly less respectable aspects of the festival.

On another occasion I was speaking with a woman of about the same age, named Patricia. We had been discussing Guy Fawkes Night when she brought up the Twelfth of July, reminded of it by her experiences in Wexford and Oxford, England. She had been living there for six years, although she was born and raised in County Antrim, in Northern Ireland. "It's a lot like Guy Fawkes, isn't it?" she asked. "It's sort of the same thing, our version of Guy Fawkes." When I asked why, she mentioned that Guy Fawkes Night is essentially anti-papist and anti-Catholic, and so is the Twelfth of July. "Most Catholics try to get away from it," she said. She repeated the idea that the bands harassed the Catholic churches. "I used to watch them gather at the top of the road and come down the Brunswick Road. When they got to St. Columcille's they all started beating the drums like crazy." Having said that, she allowed that she used to go to listen to the bands with her grandfather, who was, surprisingly, a British unionist. This is surprising because Patricia is Roman Catholic, and we met her at the "mums and tots" group at St. Columcille's. Nevertheless, for her the Twelfth is entirely a Protestant celebration. At some bonfires, she pointed out, effigies of the pope are burned.

She mentioned that friendships with her Protestant friends become strained both in advance of and during the Twelfth, but go back to normal some weeks later. Also, she mentioned that her Catholic father used to own a sweet shop in Bangor, where he would fly a union flag on the Twelfth to get the band business—"which he got in droves," she says. Here we see an example of an interesting phenomenon: a member of the subaltern population tricking the dominant group by appearing to accept and subscribe to its cultural symbols. In reality this man saw an opportunity to turn the other group's flag to his own advantage. By flying the flag he remains open and does lucrative business, gladly taking the money of those whose actions support the ascendancy of that flag. As for genuine Catholic participation in the Twelfth in earlier times, Patricia feels, like Maureen, that this was not common. The frequently cited claims to the contrary, she feels, may be simply what Protestants would like to believe.

So the parades are festive and sensually pleasurable, capable of being enjoyed on that basis by anyone. Certainly not all parade participants, nor all members of the Orange Order, march to flaunt their power in a territorial or triumphalist way, but many do. Not every Catholic reads the parades in this manner, but many do.

## Murals

Likewise, when one finds streets and buildings coded by curbs painted in the colors of one of the two opposing flags, this can be explained as decoration for a summer festival, or "just part of our tradition," but it also tells people where they are. Depending on the individual, that message might be welcome or unwelcome. The same is true of the many outdoor murals found throughout Northern Ireland. While these are less dependent on calendared events such as the Twelfth of July, many murals are painted, repainted, or touched up at this time of year, especially those depicting King William. Murals vary in their subject matter, in both Protestant and Catholic communities. Some feature generic paramilitary figures wearing masks and holding automatic weapons. While a King William does not denote what a paramilitary soldier denotes (nor does a Blessed Virgin Mary denote what a paramilitary figure denotes), the presence of even the less

Figure 5.   Republican Paramilitary Figure

threatening images tells a person that the neighborhood is unionist or nationalist, while the more violent images suggest loyalist or republican affinities. The painted environment signifies the ethnic and political makeup of a neighborhood or town, and may serve as an announcement that one is entering such a neighborhood.

Sociologist Bill Rolston has written extensively on the Ulster murals; art historian Belinda Loftus wrote two books, both named *Mirrors,* that deal with symbolism in Northern Ireland generally; and sociologist Neil Jarman has studied the parades extensively, along with the murals. According to Rolston, the first murals appeared in the Protestant areas and featured King William, popularly known as King Billy. The earliest is dated about 1920. Catholics did not take up this practice until the late 1960s, and it is said that the nationalist and republican murals show artistic conceptions and styles very different from their unionist and loyalist counterparts (Rolston 1991, 1992; Jarman 1997). In the Protestant areas, a generational shift in political imagery is evident in the appearance and increased usage of threatening paramilitary figures rather than the more demure "King Billy." According to both Rolston and Jarman, these reflect a generation of younger Ulster Protestants fed up with what they perceive as a stodgy and stifling Orange Order. The militaristic imagery is a rebellious statement to the Orange Order itself, as well as a threat to Catholic nationalists and republicans.

Loftus has said that unionist murals (or "gable-end paintings") tend to be consistently heraldic in their style, presenting assemblages of discrete symbols and symbolic objects, whereas Catholic murals vary greatly in the style of their execution and subject matter. They utilize influences from comic strips, advertising, Celtic revival art, and other sources. Although this has been true in the past, a series of very striking loyalist murals were painted along the lower Newtownards Road in Belfast during the 1990s. These murals continued the stylistic rendition of discrete symbolic subjects such as the bible, images of soldiers, and a very large paramilitary figure juxtaposed with messages such as: "Our message to the Irish is simple: Hands off Ulster. The Ulster conflict is about nationality," and "LPOW [Loyalist Prisoners of War]: Their Only Crime Is Loyalty." Like most murals in the North, they combine the symbolic or semiotic with the semantic—that is, deeply resonant images combined with highly charged words.

The Newtownards Road murals break from unionist tradition in their boldness of execution and in their design. Four gable-ends in a single housing estate are utilized, and their façade is connected by a brick wall. The muralists took advantage of these buildings to create a single work of art. The four gable-end walls and the connecting walls are utilized and all are clearly related. So the concept of assemblage usually found in a single mural

Figure 6. "Ulster's Freedom Corner"

has here been extended to include adjacent, connected murals as well. The image here of a paramilitary soldier is the largest such image I have seen to date in Northern Ireland; its overt warnings to the "Irish" are unusually forthright, as is the statement of political analysis: The conflict is about nationality; prisoners are guilty only of loyalty to the queen.

These sentiments concerning loyalty are widely shared among the Ulster Protestants, even if endorsing violence is not. When I asked two men who were working on refurbishing these murals if the local residents approved of them, one man answered simply, "Some do; some don't." But as for being British and having that held against one in the court of international opinion, there is widespread agreement. "We were born here. We are told that we are British and that we are part of the United Kingdom. We are raised to recognize the Queen and that the union flag is the flag of our country. So we salute the flag and honor the Queen and then we're told that we're criminals."

Their only crime is loyalty, indeed. For these reasons, many Northern Irish British, or Ulster Protestants (who became known as "Scotch-Irish" in the United States) feel betrayed by the British government. The Newtownards Road murals represent a dramatic break with the propriety of the Orange tradition—there is no King Billy to be seen in the entire series of murals, and the paramilitary figure, which the Orange Order denounces, dominates a gable-end. Moreover, the paramilitary figure—representing the Ulster Defense Association, the Ulster Freedom Fighters, the Ulster Volunteer Association, and other groups also referenced in the murals—is identified with previous generations of legal military units: the Ulster Defense Regiment and the (notorious to Roman Catholics) B-Specials.

One wall, however, is devoted to the figure of Cuchullain, with the legend "Ulster's ancient defender against the Irish." It is in this context that these murals depart most dramatically from all previous unionist and loyalist murals. Not only is Cuchullain represented, but the representation is of the famous statue of Cuchullain by Oliver Sheppard. This statue, although commissioned earlier than the Easter Rising, stands in the General Post Office in Dublin as a monument to those who took part in the 1916 rebellion against the British. In other words, this loyalist mural has claimed for its side one of the single most central symbols of the nationalist and republican movement.

Known in ancient sagas as the Hound of Ulster, Cuchullain has long been cast as an Irish epic hero. By extension, he has been assumed to have been Celtic, and since nationalists claim all things Celtic, Cuchullain and

Figure 7. Cuchullain as Loyalist

his feats are claimed and used symbolically as the cultural property of the nationalist movement. Thus his statue, influenced by figures of the dying Christ, is seen as appropriate for commemorating the leaders of the Easter Rising who were executed for their actions. By all accounts the Rising itself did not enjoy widespread support, but the execution of its leaders was very unpopular among the Irish. Loftus notes the many ways the image of Cuchullain has entered folk and popular expression, including not only wall paintings but also trophies and even a butter sculpture (1988:18). He was also an important subject in the writing of William Butler Yeats.

To poach this image for a loyalist mural is a bold move, one that depends on the ideology of Ian Adamson and his elaborations on theories of the Cruthin, a pre-Celtic civilization in the north of what is today Ireland. Unionists maintain that theirs is a pre-Celtic culture, and that Cuchullain was defending Ulster from Irish-Celtic invaders from the south—just as the loyalist paramilitary groups today are doing. This is really the important point: Just as the paramilitaries are legitimized by reference to previous military units, Cuchullain is by the same rhetorical strategy constructed in the image of the contemporary loyalist, fighting for his non-Gaelicized country and culture, his non-Gaelicized territory. The contestation of the land itself—is this dirt Irish or British?—is extended to a mythic hero—was Cuchullain Celtic or Cruthin? Is he yours or ours?

These four murals, bold in both style and content, are named together "Ulster's Freedom Corner." They reflect a growing impatience and militancy among a younger generation of Ulster Protestants. They are also an example of the extent to which contestation in Northern Ireland occurs in many different domains and on many different levels. The frequency of public symbolic display is due in part to the ongoing need for such displays in a context of contestation; in such a context, symbols are not merely displayed or enacted, they are used: to assert territoriality and identity, to welcome or warn, and frequently, to offend. It is often observed that flags are more likely to be displayed in areas where there are significant numbers of a population who will be offended by their presence, for such flags or other decorations are meant as an "up-your-nose" gesture of disdain and hegemonic power. As Rolston has noted, here art is a weapon (1991:69 and passim). Murals combine specific verbal messages about nationality and territoriality with polysemous visual messages.

Ironically, a painting of Sheppard's figure of Cuchullain is found on a mural in a Catholic area of Belfast that commemorates the martyrs of the Easter Rising. So we can find in Belfast two diametrically opposed uses of the same semimythical figure, one who has been used in many different contexts but always, up until now, as a figure representing Celticism, Irishness, and nationalism. The loyalist mural subverts this history of

Figure 8.   Cuchullain as Republican

display and puts it in the service of a different ideology, a differently constructed history.

Other nationalist murals show the range of styles and brightness of color that Rolston refers to. Unlike the unionist paintings, these are dated to the late 1960s at the earliest, but they have, like the unionist paintings, gone through their cycles of boom and bust, generally corresponding to periods of social upheaval, widespread discontent, or unpopular political decrees.

Murals in Catholic neighborhoods are notable for their diversity. For instance, in an attempt to establish solidarity with oppressed peoples and revolutionary movements internationally, Belfast has boasted murals dedicated to Nelson Mandela, American Indians ("Their struggle is our struggle"), and—in a mural copied from the album jacket graphics of a Bob Marley record—Jamaican Rastafarians. The images are freely adapted from publicly available sources and are often recognizable as such. When I first saw Gerard Kelly's neo-Celtic mural of the Tuatha de Danaan, for instance, I took it to be inspired by comic book and commercial illustrator Barry Windsor-Smith's 1970s art nouveau-style renderings of Conan the Barbarian. In fact, it turns out to have been heavily influenced by the posters of contemporary illustrator Jim Fitzpatrick. In turn, Fitzpatrick

Figure 9. Commemorating the Easter Rising

Figure 10. Paramilitary Memorial Wall

derives his neo-Celtic stylings from nineteenth-century art nouveau prints. So, as Loftus points out, there is a fluid and dynamic—and popular—use of artistic traditions going on here that challenges academic categorizations of "Art" and renders them meaningless.

Kelly is perhaps the most celebrated of the contemporary republican muralists. His work exemplifies the variety of styles found in the nationalist and republican areas. His tribute to the Easter Rising features the date 1916 as the central image, in red numbers writ large across a field of green. This mural is perhaps the closest I have seen to the hip-hop style of wall painting found in the United States, largely among African Americans, Latinos, and other ethnic groups. The latter paintings use the stylized rendition of a name as the principal subject of the painting; they have been related to advertising logos by at least one scholar (Austin 1996:273–275). Likewise, in the 1916 mural the numbers themselves are the primary subject of the painting.

Another important Kelly mural pays tribute to four fallen IRA soldiers. In it, the deceased individuals are named and realistically portrayed, dressed in their miltary fatigues, brandishing automatic weapons. These kinds of memorial walls are found often enough, but two points are worth mentioning. First, Kelly's mural also features a Celtic cross as a central motif, dividing the area of the painting into four parts, one for each of the individuals being commemorated. According to Rolston (1991:104–105), this is a unique combination of the republican movement with Celtic symbolism and, by extension, history and mythology. The chain of associations I have delineated above are visible here: militant republicanism today is part of a racial (Celtic) heritage; the Celtic culture of old is the Catholic Irish society of today; and the IRA is the source of its contemporary warriors.

Second, since I mentioned the American murals earlier, I will note that in the United States, murals are frequently used as memorials to fallen gang members, as well as to innocent victims of police brutality or drive-by shootings. In Northern Ireland, this kind of commemoration usually takes the form of spontaneous shrines to the victims, as will be described below. There are memorial walls to paramilitary figures, but these have the air of more formal commemoratives such as statues (see, for instance, Gillis 1994), and they serve, in a way, as recruitment posters. The deceased are presented as fallen soldiers, martyrs, heroes—but not as family members, as sons, daughters, fathers, mothers. These walls represent the political and military struggle, but not the domestic suffering that such deaths cause.

When I met Gerard Kelly, he repeatedly mentioned that he had served a five-year prison sentence. Finally I asked him why. "Attempted car bomb," he told me. "Attempted—That means it didn't go off?" I asked. Fixing me with a hard stare, he replied, "Unfortunately, it didn't go off. *Unfortunately,*" he emphasized. I asked if precautions were taken when planting a car bomb to ensure that innocent people would not be injured, precautions such as watching pedestrian traffic. "I can't tell you that," he said. "That's a military secret. I'll tell you this much. I was informed upon. The person who informed on me was found dead with six bullets in him." I shuddered as he pointed to the residence of the alleged informer. "Look," he said, "I wasn't a bad kid in school. I didn't get in trouble with the police. This is not about that. This is war."

Kelly learned his art while in prison, where he would receive blueprints for mural designs. I have been told that these murals must meet with the approval of the IRA high command. He is proud to point out that he had no art training prior to his prison sentence. "They say prison will either make you or break you," he said. "I'm hardly broken." He demonstrates no remorse as he paints the walls of Belfast's republican neighborhoods, frequently with the aid and apprenticeship of neighborhood children.

As we survey a number of activities throughout this book, we will see ways in which the children of Ulster are socialized through these activities into the "us versus them" mentality that is prevalent. As with building bonfires and painting curbs, one can see how aiding the mural artist is great fun for youngsters, and in the case of Gerard Kelly, how the individual's stature is enhanced by his personal history as well as his talent. His politics become their politics as he heroically paints republicanism on the walls of West Belfast. In doing so, he has faced danger and adversity: among other things, he claims that some soldiers once tried to kill him. He was saved only by the approach of passersby. While I was with him in a car, there were times when he hid from the view of British soldiers. Ironically, after the ceasefires of 1994, he has had gallery showings in Dublin of photographic reproductions of his work.

Gerard Kelly represents one way in which art and politics are one: The artist is the warrior. Not simply because of the prison history of this individual, but also because the murals are far more than just pretty pictures. While they may not be overt calls to arms, nor are they all advertisements for Sinn Fein, they all present deeply coded cultural symbols as part of a structured political message. Furthermore, the very existence of these paintings is seen and felt as an act of resistance and opposition; this is the perception of those in power and the ruling elite as well as of those who paint the murals and those who feel they reflect their situation. This is true of the loyalist murals as well. While shooting a gun, launching a missile, and detonating a bomb are certainly the most extreme and egregious forms of sectarian activities, one also fights the battle with wall paintings (see, for instance Peteet 1996) and many other public markings as well.

The murals are the most obvious aspects of the painted landscape; they are joined by the painted curbs and any number of related variations. For instance, in the foothills of the Mourne Mountains, outside the village of Annalong there is an ancient standing stone that is thought to be megalithic. We do not really know what function it served, although it seems to line up with others miles away. Perhaps it is astronomical, perhaps religious, perhaps both. Whatever the original purposes of these stones, today they carry an air of great mystery due to their antiquity and inscrutability. But this particular stone has been painted the colors of the Irish Tricolour. Periodically, attempts are made to destroy it with a bomb. The stone, already evocative of great mystery, is yanked out of its timelessness and fashioned into yet another, very context-bound chess piece in the apparently endless tit-for-tat of conflict, real and symbolic, in Northern Ireland.

It must also be said that there is a continuum from graffiti to mural. As we have seen, most if not all murals combine the semantic (words) with the semiotic (pictures), and the words are more or less permanent. In addition, many public surfaces are covered with graffiti, frequently of a political or sectarian nature. For instance, on a wall next to the site of an Eleventh night bonfire in Bangor is written the words "Ulster needs ethnic cleansing." Frequently the graffiti is informed by space. Written on a wall at the boundary of the Irish Republic—though in Northern Ireland—are the words, "We will not forsake the blue skies of Ulster for the grey misty clouds of an Irish Republic." Other specifically spatially related markings include the figure of the Blessed Virgin Mary as one enters a Catholic neighborhood, or the use by both sides of a threatening paramilitary figure. And of course the painted curbs mark territory visually.

One of the most striking manifestations of division in this regard is found in Londonderry, a city whose majority population is Catholic, and where experiments with power sharing in the city council have been deemed successful. The city boasts perhaps the most successful city-wide, city-sponsored Halloween festival in Northern Ireland. However, it is also a city in which a decisive moment in the Williamite wars occurred, as William of Orange broke a siege of the city by the forces of King James in 1690. We will return to this later, when we examine the annual burning of the effigy of the then governor Robert Lundy, but it is important to note here that the relief of Derry (as the city is familiarly called, with important exceptions) is celebrated twice a year: once in August and once in December.

As one enters the Protestant neighborhood, one is greeted with a gable end that announces "Londonderry/Still Under Siege/No Surrender." Just as the Catholic forces of King James had the city surrounded, so are the members of the minority Protestant population today surrounded—and "under siege"—by the town's majority Catholic population. And just as the lesson of 1690 was that by refusing to surrender they would eventually prevail, so today does this population refuse to surrender what they see as their rights,

Figure 11.  "Londonderry...No Surrender"

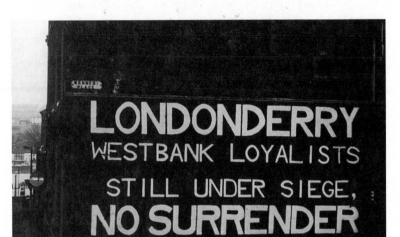

their very heritage, in the face of the Anglo-Irish Agreement, the Downing Street Declaration, and the continued negotiations between Great Britain and the Republic of Ireland. "No Surrender" and "Ulster Says No" are two very common slogans found throughout Northern Ireland, painted initially in response to the Anglo-Irish Agreement and then extended to include most of what has developed since then; but like so much of the public rhetoric and iconography of Northern Ireland, these slogans are rooted metaphorically and rhetorically in historical events.

However, as one enters the Catholic section of the city known as the Bogside, one sees a gable end that declares "You Are Now Entering Free Derry." Note the rhetorical strategies of these two signs, and the ways in which they manifest the deep contestations that occur in Northern Ireland: This is Derry, not Londonderry. The Irish name of the town, Doire, is generally accepted as predating the English presence there. Formerly in County Coleraine, as the province fell under the jurisdiction of the London Company, the name of the county was changed to Londonderry, as was that of the city. Thus to refer to the city as Derry in this context is to deny the English claim to domination, and to imply that it is "free" of British imperialism. Yet the other sign insists that this is *London*derry and that it is still

Figure 12. "Free Derry"

under siege, though the implicit message is that with endurance will come victory. We see here claims for two different realities, both having to do with the essential nature of the place: It is Irish or it is British. The two positions are more than incompatible; they are mutually exclusive. This mutual exclusivity of worldviews is at the heart of the conflicts in Northern Ireland.

# CHAPTER THREE

# Assemblage

One summer evening I drove through the town of Milltown in North Down. High above me, affixed to street lights, were a series of heraldic shields, coats-of-arms bearing such familiar but contentious symbols as the Scottish St. Andrew's Cross and the Ulster flag. The religious and ethnic makeup of the town, and its political allegiances, were being expressed in no uncertain terms to those who approached. These shields were along a major motorway, not a neighborhood street. Pedestrians were unlikely to use it at all. The shields seemed to be a statement to the passing world, a proud assertion of identity, but they certainly caused some consternation among those they excluded—the Irish nationalist population who are a distinct minority in North Down. Public display reflects many intentions and provokes a number of perceptions, but the exclusion of any Catholic, "Irish," or "Celtic" symbols—along with the semiotic compatibility of the symbols chosen—certainly sends out simultaneous messages of uniformity and exclusion. The very fact that they are so public, positioned as they are along a major thoroughfare, also speaks volumes. The messages appear to be officially sanctioned. Were these shields set up for parade routes? They are perceived by nationalists as hostile statements, and by unionists—perhaps somewhat disingenuously—as mere expressions of identity. Both sides consider the legitimacy of their perceptions to be self-evident, and paradoxically they are both correct in their assertions.

Likewise, on another summer night, when I was on my way to a Lambeg drum contest, I drove through (or under) an Orange arch erected for the Twelfth of July. This particular arch was situated at the entrance to town as one approached from the north. It was a striking arch; by my non-native standards, it was one of the handsomest. It was striking not only aesthetically, however, but also in its stark presence, its materiality, its visuality. It is a solid statement to anyone who is approaching the town; one has to drive through it to get anywhere else with any reasonable convenience. Passing

Figure 13. Arch Erected for the Twelfth of July

through an arch is a ritual filled with meaning in Northern Ireland. The Orange Order regularly parade through them, and Catholics have been forced to walk through them as a kind of ritual humiliation. The placement of the arches insures that all traffic is symbolically initiated into the unionist ideology, at least apparently, regardless of whether the individual driver cares to be or not. The shields and the arches are material components of the marching season centered on the Twelfth of July and are thus seasonal and temporary. But they are also statements that are impossible to ignore regarding the nature of the territory one is entering.

These examples demonstrate that where these objects are placed is an important element in their impact. Not only the content of the artifact but also its location, what other objects are in its proximity, and where they are placed vis-à-vis each other, are both aesthetic and political considerations. These spatial relationships combine to create visual statements that communicate larger, somewhat more complex social and political messages. This is a principle of public display that I refer to as folk assemblage (Santino 1986, 1992b), the popular presentation of discrete elements as a larger whole. Assemblage involves the juxtaposition of elements that can be and often are displayed as discrete units in order to modify, strengthen, or otherwise

Figure 14.  Union Flag with King William of Orange

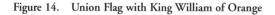

develop a symbolic public statement. Assemblage can be writ large or small. For instance, during parades, a drummer will frequently affix a bouquet of orange lilies to a Lambeg drum in representation of William of Orange. The drums are already painted with iconographic scenes of importance to the Orange Order. Thus the presence of the lilies represents an intensification of an object that is visually designated as Protestant and unionist; that is carried in the chief celebration of that ideology; and that many people perceive to be menacing in the quality of its sound. The sound alone is felt by nationalists to carry these associations, but the drums are painted with potent images, adorned with symbolic flowers, and carried during parades said to be triumphalist in nature. An assemblage such as the decorated drum thus creates a dialogue of semiotic messages within itself to create a symbolic form greater than the sum of its parts. Moreover, while the drum is itself an assemblage of several visual elements (the instrument, the images on it, and the flowers attached to it), when carried in a parade it becomes a component of a larger assemblage of other elements: banners, bands, flags, uniforms, and so on.

As I suggested above, we are dealing with the slip between intentions and perceptions. In July 1994 I encountered another example of assemblage, one in which the perceiver definitely experienced the juxtaposition as hostile, when I met Joe McMullen in Bangor. He was sitting on his front step as I took a photograph of a union flag with William of Orange depicted on it flying from the house that adjoined his. He invited me to sit and talk with him, and then proceeded to tell me about his neighbor.

He explained that because it was the summer, the marching season, tensions were high. People who were normally friendly stopped talking to each other, even in the workplace. His neighbor did not talk to him at all—Joe is Roman Catholic—and of course this neighbor flies that flag. Worse, Joe said, sometimes he attaches orange lilies to the flagpole. Joe saw this as antagonism, pure and simple; so troublesome was this flag and flowers assemblage that Joe believes it contributed to the death of his wife eight years previously. Although the cause of the death was Parkinson's Disease, Joe said that the flag with the lilies attached to the pole had bothered her terribly. He believes that his neighbor has been intentionally provocative.

Here the simple addition of the flowers to the flag compounds the impact of the symbolic statement. To Joe, the fact that his neighbor went to the trouble of adding the flowers on the morning of the Twelfth means that he went out of his way to add insult to injury. And while the injury is not tangible, Joe feared it might be. During another conversation, he repeatedly expressed his fear about talking of such matters, and said that the best thing a Catholic could do in the face of intimidation was to do nothing:

J.M.: No, now that flag doesn't bother me. But they're only doing that to live up to tradition and to keep Catholics annoyed ... When it's put out there, when [the flags] are out they're letting you know that they're the boss and we know they're doing that to try and annoy us, but they won't annoy us ...

They put the lilies on them years ago and tied them up and got out there at nine o'clock in the morning and tried to annoy you. But you'd have no chance. You're a coward. You have to be a coward. Better a coward. It's better to be a coward than a dead man for life. They'd shoot you as fast as they could, they would ... Orange lilies are flowers. You know, a flower. They had a lily, a flower at that particular time [the Battle of the Boyne] and they took it as an emblem of their fighting. They brought it and had it in front of their Lambegs and their guns. There is a song, the "Bloody Orange Lily," you hear about it. There's a song all about it. And they would hang those sorts of things. And I remember. I remember them talking about it whenever they knew I was listening. I didn't want to hear about it and they knew I wasn't fearing them talking about it.

They put arches up. You know what arches are? You've seen it up there in some places in Bangor, didn't you?
J.S.: Yes.
J.M.: Oh, Catholics, Catholics were sometimes forced through that arch, to walk through it.

J.S.: Forced?

J.M.: Yeah, you were forced.

J.S.: How do you mean "forced"?

J.M.: Well, they weren't, they weren't asked but some neighbors would come and say, "There's Joe. Let's take him through the arch," or something like that. And then there on the Eleventh night, the day before the Twelfth they'd put all the bunting up. They'd put all the bunting across the road. And tie them onto your door and onto someone else's door. And just carry on. Oh, they're silly people. Silly. There's no bloody, no bloody, there's no reason—it makes me curse and swear a bit and I don't want to curse and swear a bit.

It comes at a bad time. We stay in the house, you see, during the Twelfth and the Thirteenth we never—everyone stays in the house. And *they* all march. They all congregate from all over the six counties, the six counties of Ulster. Did you know that? Well, they all gather and go maybe to Newtownards, maybe Saintfield, or maybe they decide some other place, maybe Downpatrick or something.

Well, that's what they do and we stay at home. I've never seen a march in my life of Catholics, now, they do march on the Sixteenth of August or sometime, don't they?

J.S.: Oh, you've never seen one?

J.M.: I've never seen one. I've never seen one.

J.S.: Well, would you go to such a parade if—

J.M.: No, oh no. I won't try to annoy anyone at all. No. There is [lowering his voice] there is a risk to Catholics on them, I honestly do think, without being a bigot.

J.S.: No one would ever say anything? No one would say to the people, "Look, that flag is offensive to us"?

J.M.: Oh no, of course not. You can't.

J.S.: You really believe you would get killed?

J.M.: Oh yes.

J.S.: Even in a neighborhood like this? How long have you lived in this house?

J.M.: I'm 62 years in this house alone.

J.S.: And you think this house is bad enough that you would be afraid to say—

J.M.: If I put a flag out and it was a Union Jack I'd be the talk of the town and I'd be asked by a Protestant what I was doing with that and I'd probably get the house burned for making a mockery of their flag. ... For I shouldn't be flying the flag. I should be doing what I'm doing—doing nothing.

Throughout this testimony we can see Mr. McMullen's reluctance to speak out due to his fear of retaliation. He also recounts examples of how symbolic artifacts, on one level celebratory and festive, are used in hostile and confrontational ways: Being forced to walk through an Orange arch, for instance, is a particularly humiliating parody of an initiation rite. The fact that civil and social power is in the hands of the Protestant majority is made dramatically clear in Joe's description of his ordeal as a member of the other faith (who in fact is highly skeptical of a number of Roman Catholic doctrines, such as the virgin birth) being forced against his will to pass through the threshold of Orangism.

It is not enough that a large minority of the population has to endure the widespread public display of symbols of an historic oppression. In addition, one is made to partake in their rituals. Citing Saxl and Wittkower (1948), Jarman says that "Persons passing under the arch supposedly shared in the virtues and qualities which decorated it … In the 16th and 17th century arches were used extensively in royal pageantry in England" for triumphal purposes, an idea that derived from ancient Rome (1997:48). Moreover, the earliest reference to a triumphal arch in Ireland was to a Londonderry arch in honor of King William. The arch, then, is a public construction deeply associated with victorious military figures like King William. Parading under one identifies the paraders as sharing in the values associated with this symbolism; being forced to walk under one is a dramatization of the power of the opposition: personal, social, and political.

Similarly, red, white, and blue bunting and pennants displayed in celebration of the British state may be attached to the front door of a person who does not identify with that state. Notice the aggressive use of symbols Mr. McMullen describes: If he displayed a British flag, he fears his house would be burnt to the ground (even if he truly preferred to retain the union with Great Britain, as many middle class Catholics do). However, unionist pranksters who do not allow him to fly the British colors may themselves impose those colors on him (just as they might force him to walk through the arch). The choice is simply not his when it comes to things British.

Additionally, Mr. McMullen perceives the apparently less intimidating act of flying flags, especially when adorned with orange lilies, as primarily intended to "annoy" him and other Catholics. During one conversation he blamed the deterioration of his wife's health specifically on the annoyance caused her by the flowers and flag. Protestants generally respond that their display events are merely their own proud traditions and are not intended to intimidate. While this is clearly not always the case, nevertheless it is frequently true. The display and the ritual festivities can serve multiple purposes simultaneously, and particular uses are specific to each situation. Moreover, although intentions and perceptions may rarely coincide in Northern Ireland when dealing with these kinds of materials, this does not mean that either intention or perception is "wrong." Both are vitally important social realities.

I do not intend this book to be a one-sided condemnation of a group or political position. For instance, a member of the Apprentice Boys, a strongly unionist fraternal association, told me that while Roman Catholics might pretend to be friendly, the fact that they vote for Sinn Fein indicates that they really want to murder Protestants. Despite the fact that Sinn Fein receives only 10 percent of the overall vote, the party receives approximately 30 percent of the Roman Catholic vote. Protestants see this as proof of the violent nature of the Catholic population generally. On the other hand, members of the Ancient Order of Hibernians told me that the Orange Order was inherently proviolence. When I challenged this, they allowed that it might be an extreme faction of the Orange Order. These AOH members saw the Orange Order's parades themselves as being supportive of the violence of those extremists. Both groups, then, are utterly convinced of each other's inherently violent nature. Once again, we see that perceptions—as much as, if not more than, intentions—are integral to the situation. Indeed it is my thesis that the "tit-for-tat" mentality, each side claiming to be responding to initiatory aggressive acts of the other, renders either side "right" from its own perspective; and that in spite of the tremendous amount of emotional involvement the people of Northern Ireland have in the events and manifestations of the Troubles, their positions are rational from their own perspectives. Joe McMullen, for instance, is conflicted about many of the tenets of his Catholicism, he is the product of a mixed Catholic–Protestant marriage, and he condemns the IRA. Yet as a social being categorized "Roman Catholic" he has experienced not only the symbolic aggression described above, but has also endured threats to his life:

> J.M.: I got a job. I had been out for a job here for building the football grounds. And I got on this, I think, 35,000 pounds a year job. I got word right away. Eugene, my son, comes up and says, "Here's the matter, Daddy," he says. "You're going to be shot." I turned and looked at the darkness. And he told me it was the loyalists. They're loyal to the crown, you see. That if I didn't leave, if I was ever seen around there again, I'd be shot stone dead. And don't go back again. I had to jack my whole work up and give it to them. And they took it over. They had all the people come up from the [Protestant area of the] Shankhill.

On a wall at the bottom of the street on which Joe McMullen lives it is written "ULSTER NEEDS ETHNIC CLEANSING" and "ALL TAIGS MUST DIE." "Taigs" is an epithet for Catholics. According to Mr. McMullen, when the shopkeeper whose wall it is was asked why he didn't remove it, he replied that he would have nothing to do with it—he did not put it up and he would not erase it. In other words he did not want to get involved because of the potential retaliation. He feared that the authors would "fire something in through the window." This wall abuts the field where the Eleventh night bonfire is burned, and next to it someone had attached a union flag to a telephone

Figure 15. Loyalist Graffiti

pole. Again, it is largely the positioning of the elements that determines what they mean in relation to each other. The graffiti and the flag frame the bonfire in a very specific way. For example, a British flag by itself is by no means representative of such racist ideologies; but placed where it is here, the British flag is in dialogue with and comments on the other elements. Moreover, people who are used to encountering a general symbol such as the British union flag in such hostile contexts remain suspicious of it in other, less loaded contexts. Mr. McMullen and others too must pass this corner in their daily comings and goings, also passing nearby houses adorned with as many as seven flags. The individual in the house of seven flags may not agree with the genocidal sentiments expressed on the wall, but Mr. McMullen experiences both, and other items as well, together. One can understand how he might think that the flags flown so near the wall are in some way an endorsement of the wall's graffiti. The assemblage imprints the respective associations of the individual components on each of the others; and further, each encounter with the discrete elements may recapitulate the previous encounters.

Similarly, when I attended the 1994 Twelfth of July parade in Newtownards, I could not help but notice that there was a vacant lot among the row of shops on what is called the "high" street (the main street).

A sign explained that, "due to the recent bombing," customers would have to take their business to another branch. An IRA bomb had demolished the building and caused minor injuries to people in adjoining buildings. So when the Orange Order paraded through the town, past painted curbs and under bunting, while drummers beat their mighty Lambegs, the destroyed building provided a context for the symbols. Implicitly it was understood: This is what the IRA does to us, this is the terrorism and tyranny that we oppose. Both sides feel they have just cause and demonstrable reasons for their positions, their beliefs, and their expressive (if not paramilitaristic) actions.

We can look at whole towns or neighborhoods as assemblages writ large. As one approaches a town and sees some combination of an arch, painted curbstones, murals, graffiti, flags, and bunting, one views these as a totality. Each element can be and sometimes is displayed individually, but often they are found in combination. As such, the particular symbols relate to and modify the others in proximity. Sometimes this is intended by the displayers; at other times it is the simple result of experiencing the culturescape of a place. Thus we can examine the symbolic display both atomistically and holistically. A deconstruction of each element, and a consideration of its relationship to other artifacts nearby, can help give the researcher some idea of the impact of such displays on various members of the public, always recognizing that their are multiple spectatorships and multiple meanings: that not only is there diversity among population groups, there is also diversity within each of those groups.

## The Burning of the Lundy

Jarman also suggests that there are both temporal and spatial relationships manifest in display. He refers specifically to the juxtapositioning of images as a factor in their semiotics: "This encourages a viewer to interpret images not as self-contained statements, but rather within the wider context of adjacent and interacting images" (1997:216). I have suggested the importance of spatial relationships; temporal relationships have to do with timing throughout the year—the nature of an event in relation to those that precede and follow it. Most calendrical holidays in Northern Ireland are politicized: The marching season peaks in midsummer, but Orange parades are also associated with Easter Monday and New Year's Day. The Catholic marching season, which is pale in comparison to the Protestant, extends from St. Patrick's Day and the commemoration of the Easter Rising in March to the Feast of the Assumption of the Blessed Virgin Mary on August 15. Only Halloween is universally considered to be nonsectarian in nature by members of all religious persuasions in Northern Ireland (Santino

Figure 16. Lundy in Effigy

1996, 1998). Jarman suggests that we must look at the Twelfth as the culmination of a ritual cycle of local and regional celebrations that correspond to the time known collectively as the marching season (1997:168); and I would further suggest that ceremonial public-display events that occur throughout the year also provide contexts and produce meaning with regard to each other.

For instance, the Burning of the Lundy, also known as Lundy Day and Derry Day, takes place on the weekend closest to December 18. Each year an effigy of Robert Lundy is set afire due to his desire to come to terms with the forces of King James during the Siege of Derby in 1690. He was prevented from surrendering by the apprentice boys of the city. King William eventually broke the siege, and Lundy has become a hated figure among the city's unionist population. Since the burning of the Lundy effigy takes place in December, Christmas imagery is visible throughout the city, including Father Christmas figures soliciting money for charity. The imagery overlaps with the historical and political celebration: blood and thunder (or "kick the pope") bands may march and play music while wearing Santa Claus hats, and children may watch the burning of the effigy while wearing such a hat or eating a peppermint stick. Victor Turner has suggested that symbols combine the ideational and the sensory poles of experience; that ideology is made desirable by being associated with a positive emotional experience (1967:28–30). When I witnessed Lundy Day in Derry in 1991, while the effigy burned in the black night air of winter, a group of young men sang anti-Catholic and anti-IRA chants. Immediately next to them were mothers with young children, who seemed not to mind the strong language and sometimes shocking sentiments ("We're the boys from Londonderry/Fuck the Pope and the Virgin Mary"). Instead, the youngsters ate their candy canes, wore Santa Claus hats, and were mesmerized by the burning effigy. Clearly it was for them an engrossing, memorable experience, as they clung to both icons of Christmas and flags of separation. They were clearly enjoying their experiences, as I would have, but I could not help but be aware of the socialization that was occurring. However, the sensual delights and positive emotions of both Christmas and of bonfires in the night were conjoined with political and sectarian ideologies that were devastating to peaceful life in Northern Ireland. Indeed, they might lead to injury to the children themselves.

## Time, Space, and Conflict

A conflation of time and space occurs frequently on the murals, as when the soldiers and martyrs of previous eras are equated to the paramilitaries of today. Fallen IRA members, for instance, are regularly equated to fallen rebels of earlier centuries and to the executed perpetrators of the Easter Rising; while on the walls at "Ulster's Freedom Corner," the UVF is framed as the heirs to the original UDA and the B-Specials as "Ulster's Defenders." Jarman notes a similar phenomenon when the Battle of the Boyne is compared to the World War I battle of the Somme, in which the then legal UDA joined the British army as the 101st Division, only to be decimated at the Somme. This conflation creates "an equality of value" between events of the recent past and those of a distant, sometimes mythological past ( Jarman 1997:168).

These conflicting uses of public space have to do with symbolic struggles over national identity played out in a complex situation. The Republic of Ireland is postcolonial, while Northern Ireland is still a colony—although this might be denied by many in both Northern Ireland and Great Britain. Thus some residents of Northern Ireland identify with the postcolonial Ireland, which has been undergoing a booming economic period in the late 1990s as it enjoys the benefits of membership in the European Union. Other residents want to remain attached to the colonial power. Postcoloniality plays an odd role in this case. The struggle can be represented in many different discourses, including those of territory, nationality, and religious affiliation, but England's role as an imperial power has shaped and continues to impact the status quo of Northern Ireland today. When we think of struggle, we usually think of physical struggle or dramatic social confrontations. These are certainly present in Northern Ireland, but the struggle is equally present in the visual display of symbols such as murals and parades. The dynamics of this struggle have to do in large part with the marking of territory, or, more broadly, the marking of space to create territory. Moreover, there are two sets of struggles, both working class—one that opposes the British government in the name of being Irish; the other, in the name of remaining British.

The landscape has long been central to the ways the Irish imagine themselves and present themselves to others. Poet and essayist Seamus Heaney has written that the Irish landscape is a touchstone to cultural myth, legend, and history. One literally sees the mountains on which the young Patrick (who would become the patron saint of Ireland) tended sheep; one walks the mountains created by the goddess Maeve. The natural landscape is the physical dimension of the cultural environment, not only in terms of the alterations or the built artifacts, but also the wide range of symbolic expressions we have been describing. One looks and sees the painted surfaces, the murals, the flags, the flower shrines to the deceased, all of which form a culturescape that reflects a deep attachment of people to place—of certain people to certain places.

I think the idea of assemblage—which combines, as Joe McMullen does in his testimony, temporal and spatial relationships and components—provides an entrée into the multiple sets of narrative struggle expressed symbolically through competing visualities. Northern Ireland is a site of rupture. There has not been an accepted, official historical narrative, nor can there be one. In fact, Anthony Buckley has suggested (in a personal communication) that if one combines the officially sanctioned histories of both the Orange and the Green, one would have something that was reasonably "complete," recognizing, of course, that history is always a construct of a present that uses selected events out of an amorphous past to explain and justify that present. Histories are social and cultural artifacts no less than, say, chairs, murals, and jokes. Historical accounts of any era or any group privilege certain events according to the needs of the time in which they are written. As evidence of this, one might compare history textbooks written about the same era but produced at different times. In the United States, African Americans were essentially excluded from "history" until recent decades; more recently, women are beginning to be officially recognized as players in the historical past. In both cases, the changes have to do with political struggles that lead, often reluctantly and minimally, to historical inclusion.

The struggle over the dominant narrative is at the heart and center of Northern Ireland. The popular and folk assemblages of Northern Ireland, the referencing of objects in relation to each other (particularly in the spontaneous shrines we will examine in the following chapter) intentionally and fortuitously contribute to political meaning and to contesting narratives. The various genres and productions of visual displays refer to different constituencies, and to diversity within those constituencies.

## Social Etiquette

People in Ulster are, as a rule, cheerful, courteous, and helpful to one another. The deep political divisions of which I write, and on which the international media focuses so much attention, are avoided in daily conversation. It is considered rude to bring up issues of religious affiliation or anything that would reflect these divisions. One never asks a person if he or she is Catholic or Protestant, for instance; it is simply not done. I have had people explain to me on numerous occasions the many ways I could "sus out," or interpret codes and cues regarding people's religious affiliations. These include understanding that mentions of such things as place of residence, name or nickname, even which newspaper one reads are often coded signifiers of identity. But although there is a certain kind of silence among these people, and there are walls (literally) between them, they are in constant dialogue with each other.

The Troubles, though, are expressed through collective expressions. Tensions arise around them, but humor does as well. Tony Crowe, a member of the Apprentice Boys of Derry, spoke to me after the parade of August 13, 1995:

> There's an attitude of mind amongst most of my friends you just don't, you don't talk of it. You would very rarely talk of it. You'd say it was awful and you'd leave it at that. There's a few, you know, lunatics in every group who might—and it does crop in, sometimes the news comes on or when something happens, pretty horrendous thing, oh shit, you know, and you're aware of something, you know, like we're very aware of Greysteel, what happened at the shooting at Greysteel, the Halloween thing, and we were aware, simply

because we thought, oh shit, what's going to happen next, you know? And as individuals, I teach in a sort of fairly integrated school in what was a very Protestant area and which is no longer a Protestant area, and you're conscious of working in a situation like that. We're all, we're all vulnerable like that, like Gareth in Omagh.

But I think we all get along very well. I mean that's another aspect I must explain to you. Within a certain group, like the pub that I drink in, people don't, you know, Catholics and Protestants don't fight, you know. I mean in the sense of—I mean, there's a bit of banter. I was in last night at W.G.'s and before I left I had a fair few. Michael gives me a Tricolour and says, you know, "Carry that tomorrow, ya bastard!" Well, you don't mind that, a bit of banter. I took it and gave it to the boys for the bonfire.

This piece of testimony is revelatory for a number of reasons. It accurately reflects, I believe, the prevailing ethos that it is inappropriate to discuss divisive political matters in "mixed"—that is, Catholic and Protestant—company. But in explaining this tendency, Mr. Crowe situates such comments in terms of their potential to bring on retaliatory actions: "Gareth in Omagh" is a reference to an Apprentice Boy present at the conversation who lives in a very republican border town. It is important to keep in mind that the interviewee is an Apprentice Boy, and that the interview took place on the day of one of their most important commemorations, so his emotions were high. Still, he and the others present see no room for compromise with their Catholic neighbors, who they believe vote to kill Protestants by voting for Sinn Fein. On several occasions Mr. Crowe spoke directly or obliquely of the need to fight for one's country. Thus, when he mentions his teaching occupation in a "sort of fairly" integrated school in an area that was once "very Protestant" but is no longer so, he is taking a swipe not only at Roman Catholics but at the system of government that he believes favors Catholics at the expense of Protestants. And when he talks of a man whom he likes as an example of the banter that substitutes for confrontation, he concludes almost as an afterthought that he had the Tricolour flag burned on a bonfire.

There is dialogue in the public sphere, as Habermas understands it, of course; that is, in the pubs and through the popular print and television news programs. But I argue that the visualities I refer to—the parades, banners, bonfires, murals, and spontaneous shrines—are not only important media of discourse among sometimes overlapping but differing population groups, but that they are the principle means of such dialogue. They, even more so than the mass media, are the true popular culture of Northern Ireland. As such, they take their place with the sectarian shootings, that are but an extreme on a range of dialogic activities.

## Bonfires

Bonfires are a common component of celebration in Northern Ireland and throughout Great Britain (and of course elsewhere as well). David Cressy (1989) has demonstrated that public, civic commemorations were thought to be incomplete without the bonfire. When asked to elaborate on his comment about the bonfire, Mr. Crowe said: "Well, bonfires are a traditional expression of celebration in this country in *both* communities, and it's traditional that effigies were burnt ... They had bonfires all along here last night." At which point another man volunteered, "There was one in the Irish state as well." The latter comment refers to Protestants in border counties who were expressing solidarity with the Northern Protestants. Bonfires are associated with a great many festive occasions throughout the year, including the Twelfth of July and the Relief of Derry, but also the enactment of the internment laws, the Feast of the Assumption of the Blessed Virgin Mary, the Feast of Sts. Peter and Paul, and Halloween (see Santino 1997; Gailey 1977; cf. Newall 1972). Of these occasions, some are Protestant events, some are Catholic; some are more overtly nationalist, republican, loyalist, or unionist in terms of either what is being commemorated (for instance, bonfires in West Belfast at the time of the internment commemoration would be a largely republican phenomenon). Others, such as at Halloween, are avowedly apolitical. As I have noted elsewhere, Halloween is almost universally said to be a nonsectarian event. Yet there are elements of it that are interpreted differently by the different political constituencies; I even had one man tell me that actually it was (the British) Guy Fawkes night of November 5 that was being celebrated in Northern Ireland on October 31 (Santino 1996). Also, unpopular political figures have been known to be burned in effigy on Halloween bonfires. Regardless of these examples to the contrary, people experience Halloween as a nonsectarian festival. Still, the widespread use of bonfires in tandem with overtly sectarian and political events has cast a shadow on the Halloween bonfire. In my field research I found people reluctant to discuss them, and the Northern Irish themselves have tended to confuse the two. There was a time, a generation ago, that "Bonfire Night" meant Halloween, celebrated by everyone; but today the term refers to the Eleventh of July, celebrated only by unionists.

These Eleventh night bonfires are most certainly politicized and known for their rowdiness. I was cautioned to stay away from them. I did not, but I noticed that because I had spent time in advance of July 11 at many of the bonfire sites, and had made myself known to the young men who constructed them, on the night of the conflagrations someone would appear almost magically out of the crowd at key moments to tell me what was acceptable to photograph and what was not. Indeed, the next day some fatalities were reported at some of the fires. One man had his throat slit in what appears to have been a long-standing conflict. These murders were not sectarian in nature, however.

One such bonfire site was on Sandy Row in Belfast, a well-known loyalist area. A group of boys had been working for weeks assembling combustible materials, which by now had filled an entire vacant lot. The idea is to build the biggest bonfire one can, and if possible to destroy the efforts of others. Here the competition is intergroup, among rival groups of adolescent Protestant boys associated with each other by virtue of neighborhood. For this reason, the boys do not actually build the bonfire structure until a day or two before the Eleventh, to prevent rivals from setting it afire early. This same kind of youthful rivalry is associated with the Halloween bonfires and other customary constructions in other European traditions, such as Maypoles in Czechoslovakia, for instance. Also, boys will sleep at the bonfires overnight as a way of protecting them—as well as a way of having fun. Sometimes small huts are built adjacent to the bonfires for this purpose. At other times, the construction that will be the bonfire is itself used as temporary sleeping quarters up until the night of the fire.

On this particular July Eleventh, it was afternoon and I was talking with a group of boys who had been collecting materials, building and working at this site for weeks. All their work was now prepared for its glorious culmination later that evening. The wooden crates, planks, branches, and tires were now a huge pyramid topped with the Tricolour flag of the Irish Republic. While we were talking, a woman approached the group asking what the purpose of all this was. She was Australian and had been on a tour in the South. On her own she decided to visit the North because she had heard that this was a particularly interesting time (the July Twelfth holiday) to be there. Unfortunately she knew next to nothing about the realities of Northern Ireland and began to ask questions that were embarrassing to me and made the boys uncomfortable. "Does this have something to do with the IRA?" she asked. The boys averted their eyes while one mumbled "no." A boy who was an unofficial leader and spokesperson for the group explained the imposing wooden structure as "British tradition." "But you're Irish!" she countered, again to their embarrassment. The questions continued in this vein until she walked a little bit away to take photographs of the unlit bonfire. I decided I should tell her to be careful about what she said—I was beginning to fear for her, although not from these boys. This was not a neighborhood in which I felt she could be so comfortably ignorant. I explained that these boys and their families and friends see themselves as British, and see the IRA as the enemy of their way of life. "If they are not Irish," she demanded loudly, "then why do they have an Irish flag on top of their bonfire?" "Because they are going to burn it tonight," I told her. Up

Figure 17.   Bonfire with Irish Flag to be Burned

until that day a union flag had waved from the top of the structure, but that night they would light their fire and symbolically destroy the hated republic.

Unlike bonfires, flags are displayed outside of festival contexts. However, they are displayed in far greater number during special times of the year, and more is done to call attention to them at those times (such as attaching orange lilies to them); and they are used by local and national governments to signify festival space, being hung on public utility poles on center city streets. The very abundance of flags flown at certain times of the year calls attention to itself and marks certain dates and periods as significant. Social dramas such as the burning of flags on bonfires and the teasing of a rival by handing him a flag representing the "other side" are more likely to occur during these festive calendrical periods of heightened nationalism, tension, and license that involve traditional and religious festivals, historical commemorations, and political demonstrations. In short, flags are an important component of celebration as well as being display objects in their own right.

The same is true of murals. Like flags, they are displayed outside of festival contexts, but many of them take on a new life during festival. The towns that I have suggested we approach (in more ways than one) as assemblages writ large stand out in summer because they are given fresh coats of paint during the marching season. Festival is an inclusive genre, one that incorporates other genres: games, parades, dance, food, music, song, noise, costume, banners, flags, bonfires, and so on. Further, many of these components, such as parades, are themselves made up of other genres and components: banners, art, musical performance, marching, and so on. Of course, the banners, arches, and murals all display several symbolic objects within a single frame. Working our way back out again, we see that the elements are juxtaposed to each other (orange lilies on Lambeg drums, banners held high above flute bands); the parade passes in front of murals, under arches, past flags and bunting and perhaps some aspect of the environment that is particularly highly charged, such as the site of a recent bombing. Spatial relationships are one way of generating meaning and are an aspect of communication.

## Banners

The various properties of the objects may also be significant. Because these objects are polysemous symbols, there will be many readings of them and the actions involving them, but the symbols are themselves multivocal: they use many voices to speak to us. For example, in parade bands the musicians may be wearing uniforms of a single dominant color, as well as sunglasses. Most members will be playing flutes. This is a melody flute band, and it is seen by some as a degeneration of more musically complex bands that were more common in the past. When the band plays, its members might perform what nationalists consider a swagger and unionists will say is merely a rhythmic swaying to the music. This type of band is known colloquially as a "Kick the Pope" band, and most of them are associated with Protestant paramilitary groups. Their swagger is interpreted by Catholics as a form of triumphal gloating. Likewise, the Lambeg drums are painted with scenes from the bible of particular significance to unionists, as are the banners they carry. Buckley and Kenney have demonstrated that these banner scenes depict images of a chosen people who are threatened by a surrounding hostile people, along with images of King William and scenes from the Williamite wars (1995:175–193).

The banners exemplify the dynamics of assemblage. The images on them are often flanked with other symbols, while the banners themselves are carried in parades, surrounded by participants dressed in the uniforms of their organizations, melody flute bands, flags, the terrain of the parade route, and so on and so forth. All of these elements contribute to the overall meanings generated, and to the meanings of the particulars.

Figure 18.   Banner Held Aloft in Parade

Banners are important for all the players in the Northern Ireland public discourse; banners are so important that new ones are introduced by means of rituals of their own. The documentary film *The Thompsons* (Lawrence 1996) shows such a ceremony for an Orange banner in rural Ulster, and banner painter W. J. (John) Jordan discusses them below, when we were talking about banner occasions.

W.J.J.: A lot of parades, and even in between [major events] they get an awful lot of parades, small parades. Maybe a dedication of a banner, you know.
J.S.: Dedication of banners, did you say? So, after you paint a banner they will have some sort of ceremony for that? Would the AOH do that as well?
W.J.J.: Yes. They would mostly. It's a good way mostly to raise money, mostly for the banner divisions and LOLs [Loyal Orange Lodges] they mostly need them. Some of them, larger divisions will be pretty rich. AOH division is strong, strong division, 520 members on the post and they have got 500 part time members. It's good to see it.
J.S.: I am just sort of curious. What would they do, they would march to the church and then bring the banner in? And then what would happen? If it was the AOH would they have a priest bless it or something?
W.J.J.: Sometimes, ah, I never go to them of course, Jack, I am out on that.
J.S.: From what you know, though.

W.J.J.: It's mostly they unfurl their banner and after that there's the dedication of the banner when the priest would bless it, outdoors.

J.S.: And what would the Loyal Orange Lodges, would they do it too?

W.J.J.: Oh, they'd all do it.

J.S.: They would bring it to a church as well?

W.J.J.: Well, mostly not to a church. Sometimes now, I've had two or three this year that've just gone to churches. They've told me that they've gone to churches. Mostly I have to roll the banners from the bottom up which is very hard to do ... It's really funny stuff. Roll it up, slide the pole out and then tie three bows on the banner, and the banner will roll herself out.

J.S.: Do they do that for every new banner? I mean do they always have some ceremony?

W.J.J.: Every new banner will.

John Jordan paints banners for both the Catholic, nationalist fraternal organizations, and the Protestant, loyalist organizations, but he will not work for paramilitary groups. Still, he finds that some individuals do not appreciate the fact that he works for both sides, and so he lives with an element of fear in his life—for instance, he told me that if the doorbell rings unexpectedly, he does not allow his children to answer it. In order to avoid the issue of allegiance to a particular side, at least in part, he belongs to a maverick fundamentalist group, thus falling outside of both Roman Catholic and traditional Protestant denominations.

W.J.J.: Well, actually some of the LOLs would maybe send me a letter, and some of the AOH-ers would maybe send me a nice letter too, of thanks. Some of them would maybe put on "brother," because they think you're one of them, you know. But actually, I am not. I don't take active part, I can't. Ah, I class myself as an artist. I see it all different, from a different point of view, than a Protestant would see it all. I have an Orange Banner and an AOH banner sitting side by side. That's the only place here you don't gather the arguments or fight with them and actually, they're just banners. There's no problems with that. Ahm, some of the Protestants would say that I shouldn't do work for the AOH, which is unfair for them to say. I am an artist basically, I am not a politician. I don't take active part. I used to actually go to the AOH parades to see some of their banners, actually to get new ideas. Ah, so then when they would come to me with them, then I could give them the ideas. Ah, I would love to change their standards completely. Ah, to do new banners, different designs for them, change it around.

J.S.: Why?

W.J.J.: They've kept over the years, keeping the same old designs, you know, they've never let go of it. You see them with the LOLs and Orangemen, they would keep mostly the same pictures too, you know. So again, I try and persuade them, don't go for that, really, change it around. But the older hands, mostly the older men will insist on the replica banners.

J.S.: Why do you want them to change it around, though, just because it's more interesting to you? Or . . .

W.J.J.: It's more interesting to me and at the end of the day I feel that maybe they should, I think a change is good, it's good, it's really good and I think things should be changed around a bit. There's like two traditions in this country, and one pulls from the other and it's no different from the other, then, really. I've studied both sides and I can't make heads or tails of it really. It's all equal in my sight so I must keep them equal in my sight. Ah, some of these men maybe be bringing me a gift afterwards or something like that, and then maybe be bringing me a gift with their emblem on it, or something like that. And the Roman Catholics they would maybe send a letter of thanks or something like that. But they will not send me gifts because they feel that me maybe being a Protestant, they think that this would offend me or something, which it wouldn't. You know sometimes, I try to open out to them and let them see really, you know, what I feel. It's hard for the men to grasp after that. It's just I feel completely different, I've different views really. Ahm, that's why I work for both sides. I enjoy it, I enjoy my work. You see there's differences even between the loyalists and a Protestant. You see, Jack, I've got to stay neutral. Or else I don't survive. I can survive, but at the end of the day, I enjoy both sides. My father was the same, he enjoyed both sides of it. Ahm, it's very difficult in this country to do, its difficult, and also to bring up your children. I've got three girls. Ah, I educate them on the Scriptures ... And, ah, that's what I, more or less, I've given them. I give them a free view on it all and let them figure it out for themselves. My wife, she asked my dears there, on the Twelfth of July would they like to go to it? And they didn't want to go at all.

J.S.: Would it be unusual for a man who paints banners and so forth to paint for both sides?

W.J.J.: Yes. Actually the area artists would say that you need to watch that you don't get yourself shot or something. But, there are artists that say that.

J.S.: Other artists have said that to you?

W.J.J.: Yes, ah, but at the end of the day I've always stayed out of it. I don't fear them at all. If someone were to come here with a gun, yes I do fear it to a certain extent. Ah, but I fear it more that the wife and children get shot, not for myself. Ahm, ahm, some of the other artists, they'll advertise, a lot of them advertise in the Orange Standard. I don't advertise. If you're good at your job, Jack, if you're good at your job, you don't have to advertise. So if you try and stay neutral it's really the best, it's hard and, ah, hopefully my dears will be protected.

J.S.: You mentioned the marching season, what exactly is the marching season?

W.J.J.: The marching season will start on the 17th of March. That's the beginning of the marching season. Ah, you'll have the 17th the most I ever work, flat out, beginning this December and I'll still be working flat out come September, you know, mostly doing my scrolls, designing. I've got three arts. Or three trades actually. One would be designing, another would be, ah, sign writing, and the third would be the banner itself. Mostly for the first couple of months I'll be designing, designing my work. Ah, I've got to try and figure out the finished product. Sometimes I can get the first four out before Christmas. First four banners.

J.S.: So the marching season begins with the 17th and then what would be the events next?

W.J.J.: The 17th of March, and then after the 17th of March, you'll get really Easter time. You'll get more parades, you'd have the Junior Orange order will march, you would have the Apprentice boys coming out too. The republicans will be marching, but then I don't do work for the likes of that, I wouldn't work for the republicans. Ahm, one rule when I started up, Jack, was I would not do work for the likes of the paramilitaries. There's no way. I'd automatically chase them, I'd chase them off. If the UDA or UDF would come to my door, I would chase them off. I won't have it; it's the same with the republicans, I haven't got time for that. The AOH is mostly based on, ahm, the AOH is a God-loving Roman Catholic. I've gone to their church and also the LOLs most of them is pretty good church attenders.

J.S.: How long does a banner last?

W.J.J.: An average, it all depends on the class of silk. Ah, I use a heavy-duty silk, my silk comes from Switzerland, the silk that I use. Ah, a medium silk should last around 15 years.

J.S.: 15?

W.J.J.: And if it's a heavy-grade silk, I can't see why they can't do 25 years for a banner. Although some of the men will come along and say to me that the banners are not made of the same stuff they were years ago. But I would say to them back again, that's not really true. The only problem is that the men are more, ah, rough with my work, treat it very badly. There's not the same quality of men marching nowadays as there was years ago.

J.S.: You really think that's true?

W.J.J.: Yes. And, ahm, also they have a thing nowadays they hadn't got years ago. All sorts of things that they call an acid rain. And that's very sore on a banner. They didn't have that years ago, 25–30 years ago they didn't have that.

J.S.: So pollution hurts?

W.J.J.: Pollution destroys the work. There is a good demand but then at the same time, a lot of the men, a lot of the men is turning from banners and going to what we call these standards. They're bannerettes.

J.S.: Are they just smaller versions?

W.J.J.: Just an awful lot smaller, one man can carry it. I think the biggest banner I've ever repaired in my life was an AOH banner, and when I was painting it, it was actually, it was actually bigger than me. I've never seen anything like it in my life.

Mr. Jordan sees a decline in the quality of man who carries the banners, along with a diminution of the banners' size and a corrosion due to industrial pollution of the environment. He tries to limit his work to the nonpolitical, but his admission that the people who carry the banners are rougher with them implies that in the context of use, the banners, like the Lambeg drums, are indeed political props, if only in these contexts. The sound and history of the Lambeg drum is militaristic. Kinesthetically the marching and body movements are significant; as are the instruments chosen for performance, the tunes performed, and the style of performance. Sometimes even the materials from which the instruments are made are loaded with meaning.

Or rather, multiple meanings. John Joe Bradley, a Catholic and a member of the Ancient Order of Hibernians—a Roman Catholic, nationalist fraternal organization—told me that in fact he enjoyed hearing Lambeg drums;

J.S.: You say you like Lambeg drums?

J.J.B.: I did, but probably if I said it publicly I would be very unpopular, but I don't care, I guess. I do have an ear for them.

J.S.: What do most people think of them?

J.J.B.: Oh, that it would be the Orange side's discrimination. Well to me, it's not that, but it could be. People could take it that way, people could take it that way.

J.S.: I guess what I'm asking is, *do* people take it that way?

J.J.B.: As a matter of fact, my wife—we went out one night and we were in this one town. There was a Lambeg drum and a party going on up the street in this one joint. And the moment I heard them, I said we got to go hear these, and she would not go near them. And it's a sort of fear that was in the back of her head. She must have associated something with them and I thought she was frightened. But I went to sit for ten minutes to listen to them.

I notice in this testimony that, like Mr. McMullen, Mr. Bradley talks of how his wife, not he, was offended by the objects, sounds, and actions in question. However, he then told me that he did not and would not get out of the car while listening to the drums for safety reasons. On another occasion, a Catholic woman told me,

"Catholics will tell you 'Oh, I hate to hear it.' It's a deep throbbing that's really mysterious and aggressive, that's how they say it sounds coming across the fields. When you're there of course it's not. It's often a display of prowess. There's no doubt it varies, attitudes vary with those who march as well. They are often used for political speeches. Politicians will actually play them at meetings, and it's not just about religious themes, it is frequently about political themes."

Lambegs are perceived as threatening not only because of their loud booming sound, but because they have been overtly politicized by leading unionists such as Reverend Ian Paisley and George Molineaux of the

Orange Order, who, like Paisley, has carried them in political rallies. Also, Mr. McMullen remarked frequently that he never attended nationalist and Catholic parades because he felt it was asking for trouble. Similarly, on several occasions I have had men explain to me that they were not offended by the display of the union flag, but quite literally acted as though I was out of my head to suggest they might fly an Irish Tricolour. Catholic males frequently express a sense of intimidation when it comes to participating in their own public-display events.

Related to this is the fact that although both sides have parades, murals, flags, bonfires, and so on, it is widely accepted by people of all political backgrounds that these forms, particularly the parades, banners, and marching bands, are much more widespread among Protestants and are indeed their principle expressive forms. On the other hand, Catholics are felt to "own" Irish music, song, storytelling, and dance. While these broad generalizations are simplistic and actually untrue, there is a greater preponderance of all the activities I have been calling "public-display events" among unionists. Nationalist John Joe Bradley repeats this generally held observation, but at the same time his own testimony implicates the nationalist side in the public uses of the same forms of symbolism. For instance, first he describes an AOH parade held illegally in 1969, the time when the contemporary period of the Troubles was beginning. He professes innocence regarding any problems this might have caused, but to defy a parade ban is clearly a dramatic political act. He goes on to compare Catholic and Protestant uses of arches:

> J.J.B.: We've [the AOH] always marched. As a matter of fact, in 1969, the Minister of Foreign Affairs, that's when the British army came in and that was the Twelfth of August, the 13th of August, they came into Northern Ireland. The Minister of Foreign Affairs banned all parades on the 15th of August. But we marched together. We wouldn't stay indoors, and thought there'd be no harm. As a matter of fact some of our members were killed because of that. It was just over the hill here, that close. And all the local divisions in County Derry all went and defied the ban. The people were defying the ban. People were fined. People were sent to jail and were serving prison sentences because of it. We didn't deserve it because we didn't think we were doing any harm by doing it, although at that time the country was rather tense but we thought we were marching in a 99 percent nationalist town and we would be causing no offense to anyone. That's like with the arches and the decorations and all, well, I think it's more of the unionist side than the nationalist side.
>
> J.S.: I'm beginning to think you're right.
>
> J.J.B.: I am right! It's 100 percent stronger on their side than it is on ours. We don't put so much into them. We don't have to. We never did murals here. The Hibernians used to put up arches across the road.
>
> J.S.: The Hibernians?

J.J.B.: Yeah, used to, but that's been just dropped over the years. I think it's too much bother. I remember them in this town in my day. If there was a major parade in this town, they're put up—only they'd be put up from maybe the night before and taken down the day after. I think maybe [the Orangemen] have something to prove, but I know a lot of good Orangemen. Decent people. And I get on well with them. They know what I am and I know what they are and there's no hassles, you know.

In a complementary fashion, a gentleman who is very highly placed in the Orange Order once compared the love of a parade in Ulster to the Nazi uses of festival in Germany. He did this with absolutely no sense of irony, embarrassment, or self-consciousness. However, it must be pointed out that he extended this love of parades to the nationalist community as well, and included them in his implication. Also, he was not comparing the Orange Order to Nazis per se; I know he would despise everything the Nazis represented. Rather, he reflected a widespread sense that Ulster, being British, is therefore Teutonic, and that its people share certain expressive traits with their German "cousins." I will call this man Paul Smith; these are his words:

> We love to parade. It's probably, it's part of whatever makes us tick. You see, in terms of nationality, the Ulsterman—it's not this beautiful idea of the WASP, white Anglo-Saxon Protestant type thing. We're all mongrels, you know, we're part Scots, part Irish, part English, so on and so forth. But all that coming together in whatever sort of mixture it has, it has left us with a love of parades, a love of the color, of the music. In the sense that there is celebration, we love to celebrate in that way. Somebody told me that we're very like the Germans in that sense. The Germans love parades, whether it be down in Munich and the beer festivals and things like that, or whether it would be what you would have associated more with the Nazi aspect of things and the way they used parades. But whilst they used them for a specific purpose, there was no doubt that the people loved parades. And we do, as a people. And it's from both communities. It's not just a Protestant thing. The Roman Catholics love parades as well, because they have a stack of parades too.

This is unusual testimony, not only for the surprising comparison to Nazis but also for the defining of the "Ulsterman" as a "mongrel," and the extension of the cultural proclivity toward parades to Roman Catholic culture as well. While it reflects the widespread understanding of the British people in Ulster as "Teutonic," and thus Germanic, it also complicates that racialist tenet. For their part, a representative of the Ancient Order of Hibernians described the organization in terms that also link race and religion:

> "My father was an Hibernian for all of his lifetime. It runs in the blood, really. It's something that, when a person joins the Order, it gets a grip on

them. It's really very hard to get away from it, what it stands for, you know— it's a Catholic order. It's a symbol of your faith, and your belief. And your heritage, your history, your heritage. And your country, all combined in one."

Here, blood, history, heritage, and nation are all linked, all collapsed within a cultural tradition.

Nothing is as simple as it seems in Ulster. Still, it is necessary for all interested observers to understand that the various designations people use to refer to themselves are naturalized. They are said to be racial rather than cultural characteristics, "in the blood," and therefore justified as unchanging and unchangeable.

# CHAPTER FOUR

# Rituals of Death and Politics

On February 5, 1992, two masked gunmen walked into a crowded betting office on the Ormeau Road, in a Catholic working-class area of South Belfast, and opened fire. Five men, including two 17 year old boys, were killed. Fourteen others were injured. The attack was by a loyalist paramilitary group in apparent retaliation for an earlier IRA attack at Teebane on some contractors who were working for the British government. These are the "tit-for-tat" killings: an eye for an eye; murder answering murder.

A week after the killings, Michael McCaughan, who is a professional photographer, drove past the site of the killings and noticed that the façade of the place, Sean Graham's, was covered with flowers, newspaper clippings, and handwritten notes. Excited by what he saw, he considered taking photographs of the site. He asked me to accompany him. I was using the Ulster Folk and Transport Museum as an institutional base for conducting research on customs and festivals in Northern Ireland at the time. He described the scene to me: People had taped bouquets of flowers, newspaper obituaries, and personal messages onto the building's shuttered façade. Flowers filled the fronting sidewalk, piled so deeply as to cover and hide the doorsteps. The friends and families of the victims had spontaneously created a shrine to their memory, and the sight of it haunted Michael. After some discussion, we drove down to the Ormeau Road so I could see it for myself, and he could photograph it.

The pictures he took that day of the notes and flowers placed in front of the closed shop formed the basis of a photographic exhibit entitled *Displayed in Mortal Light*. A catalogue accompanied the exhibition, for which I wrote an essay (1992a). In this chapter, I will discuss the motivations of certain people who have created and maintained such shrines, including the Sean Graham's memorial. I will view this as an emergent tradition internationally, and locally as another one of the multiplicity of visual expressions of emotion in regard to the violence and the killing in Northern Ireland.

## Spontaneous Shrines

At first Michael was reluctant to take the pictures at all. Aware of the preda-
tory and voyeuristic dimensions of photography, he felt that he would be
violating someone's personal grief; that he did not have the right to take
these photographs. I argued that, on the contrary, he had an *obligation* to
take them, and it was because they were publicly displayed that I felt this
way. Created by a network of people who knew and loved those who were
murdered, the shrine was displayed on a heavily traveled public thoroughfare. The photographer need not intrude on anyone's personal feelings or
enter their private space. The flower shrine called attention to itself; it called
passersby to take notice of the awful events that had occurred at this place.

This shrine is neither unique nor isolated. Horrific murders throughout
Northern Ireland and elsewhere are frequently marked in such a way, along
with untimely deaths due to sudden, shocking events such as automobile
and airline crashes. The place where the death occurred is festooned with
items such as wreaths, flowers, and a wide range of personal memorabilia.
For instance, the destruction by international terrorists of the airjet over
Lockerbie, Scotland, in 1992 provoked people to fill an entire field with
flowers at the site of the crash. In the United States, flowers and wreaths
were placed on the sidewalk where a 27-year-old English woman was shot
to death by a robber in New Orleans on April 16, 1992. In fact, the practice
of creating a roadside shrine at the scene of a fatal automobile accident is a
long-standing tradition among Hispanic peoples in the United States. These
shrines and memorials, usually roadside crosses and altars, are most frequently seen in the American Southwest. The custom has spread, and today
roadside crosses and other such memorials including trees and telephone
poles tied with ribbons are found throughout the United States and internationally. They are no longer identified with a specific ethnic, regional, or
national group, although there are differences in style between one group
and another in their death rituals (see, for example, Zimmerman 1995). For
my purposes here, however, I am interested in the broad emergence of this
tradition rather than the stylistic differences.

For instance, in November 1992 an African American man in Detroit
was killed at the hands of police while in custody in an incident uncomfortably similar to the then recent police beating of Rodney King, which
sparked the riots—or rebellion, as many insist it be termed—in Los Angeles
earlier that year. A shrine strikingly similar to those I am discussing in
Northern Ireland soon marked the spot where the fatal beating took place.
Created without instigation or coaxing from any church or municipal government, these spontaneous shrines are temporary monuments to the
deceased. They are created by regular, everyday people who feel a need to

commemorate the loss of a life, to call attention to how the life was lost, and to consecrate the place where the unthinkable happened.

In the years following the attack at Sean Graham's, the incident has become emblematic of the contested, divisive nature of the Northern Irish Troubles (Douglas 1966). Parades organized by the Orange Order, the primary Protestant, unionist organization in the North, pass by the site, much to the objection of local residents. The Ormeau Road Residents Association has become one of the principal proponents of the "Reroute Sectarian Parades" movement; at the same time, the nature of Sean Graham's as a commemorative site has become formalized with the addition of a permanent plaque and annual ritual actions, such as the laying on of flowers by Irish Foreign Affairs Minister David Andrews on the sixth anniversary of the attack (*Irish News,* Feb. 6, 1998).

The plaque lists the names and ages of the five who were "murdered for their faith on 6th February 1992." It then says, "Also in memory of all other local people who have been murdered for their faith." According to this interpretation, the victims were killed because they were Roman Catholics. Since the killings were random and indiscriminate, this is accurate. It was, we can safely assume, important to the paramilitaries that Catholics should die in retaliation for the IRA's activities. Still, this indicates the confluence of religion and politics that marks the Northern Irish conflicts, and suggests that we need to understand, as Buckley and Kenney have suggested, that religion operates as ethnicity rather than as dogma in these situations, in this Northern Irish context (1995:7).

Note also the emphasis on the local, even as the site and its permanent plaque attract increasingly broader spheres of attention: throughout Ireland, throughout Great Britain, throughout the world. The initial temporary, ephemeral assemblage was created by those closest to the victims and spoke directly to those who lived and worked closely with them. But in time the site achieved an emblematic dimension in addition to its initial deeply symbolic one, although it always had a public dimension and thus spoke to such people as a passing photographer and a visiting American folklorist. In the permanentizing of the place, however, the local residents reclaim its specificity and its local importance. The devastation was local and personal, and its signification will remain so.

## Not an Unimportant Failure

Standing in front of Sean Graham's on the Ormeau Road was almost emotionally overwhelming. There is a materiality, a weight, a three-dimensionality to the shrines that photographs do not convey. The flowers on the sidewalk were consciously and artfully arranged in such a way as to allow people to enter the shrine—to approach the metal window shutter closely, to read the messages, newspaper clippings, and prayers. The shrine surrounded us and became its own environment. A large hand-drawn poster positioned centrally on the shutter featured five roses "In memory of those killed on the 5/2/92." On the poster was taped a photo clipped from a newspaper showing a woman kneeling in prayer at the door of this establishment, leaving flowers. From the newspaper photo I could see that there were two front steps to the entrance. On the day of my first visit, the flowers were piled so deep that I did not realize that there were front stairs at all.

Obituaries of the deceased had been clipped from newspapers and taped to the shutter. Handwritten notes on lined white paper torn from spiral notebooks carried messages to the deceased. The distinctive striped neckties of schoolboys' uniforms were tied in bows around bouquets of flowers in memory of the two victims who were still in their teens. The neckties would be instantly recognizable to passersby as icons of school-age boys, just one example of the ways the shrine communicates. Rosary beads were draped from flowers as well, signifying to anyone who might not be aware of it that the murdered people were Roman Catholics. Of course, everyone in Northern Ireland was aware of this, so this would not have been the primary reason for placing the beads there. But photographs of this shrine were picked up by the British press and circulated internationally. Whatever the intentions, the public display of these items ensures that the messages and meanings travel far beyond the neighborhood, and that the meanings cannot be controlled.

Along with the flowers, neckties, notes, and rosary beads, there were prayers begging God for an end to all the violence and suffering. Some were handwritten: "Peace in our time, and for our children's children/God of our fathers, hear us/ Let there be *Peace*" implored one, on ragged paper as unpretentious as the neighborhood from whence it came and the people of courage who carry on. As Michael and I stood there, boys played football around the corner in Hatfield Street—the betting office is on the corner of the Ormeau Road and Hatfield. Eventually, one of the boys joined us, silently reading the clippings and notes. Then a couple in their twenties came by, followed by another couple in their thirties. All stopped. Nothing was said. People took their time, then moved on. The youngster remained for quite a while. I saw no tears, heard no voices. The mood was one of solemnity and respect. As the children played nearby, I was struck by the fact that it was in this very spot a week earlier that innocent, average people, members of this community with wives, children, parents, friends, and lovers were killed. Such violence is part of growing up for these children. It is one thing to know that in the abstract, another thing to encounter the people and place where the killings occurred. By constructing this shrine, the people of this aggrieved community did more than pay respect to their dead. They made the rest of us take notice as well.

We all saw the reports of the murders on television, heard the stories on the radio, and read about them in the newspapers. The shrine communicates different messages than do the print and electronic media, and in different ways. Here, evil has erupted into the everyday world, and the people of the neighborhood have responded with the almost liturgical language of ritual: the flowers and wreaths that, along with the other components of the shrine, mark this spot and set it aside from its surroundings. This place, the shrine says, is no longer mundane space. Through the shrine, people declare that the building in which the murders happened is now special. The shrine jumps out of the ordinariness of the street. It forces us to confront the humanity of those who were killed and those whose lives are left behind in the wreckage. These notes on ragged-edged paper were written by real people. The neckties entwined in flowers were worn but a few days earlier by the unfortunate young men who happened to be in the wrong place at the wrong time, and were destroyed by the one great cancerous sore that eats away at this society.

Viewing the shrine brought home to me the humanity, the personhood of those who were killed. Through this display of their relationships and the devastation resulting from their loss, I encountered real people who had worked, laughed, struggled, cried, and played, and now were dead. They were killed *here,* the shrine says. They were killed *now.* They were killed ... *why?* These are their pictures, their effects, and the newly broken families they leave behind. In order to get at underlying causes we must make abstractions; but these deaths are not abstractions, and they are not mere statistics, to this community of family and friends.

Certainly the shrine is personal. The messages are from wives and children to murdered husbands and fathers. But these messages are made public for the neighborhood and for anyone else who happens to pass by. The shrine is itself the creation of the neighborhood. Many, many people contributed flowers or other tokens of their concern to it. Later I found out that a great volume of flowers had been sent up from Dublin in a show of support for those living in the North. So the components of the shrine transcended the local neighborhood, and those living in the neighborhood were aware of this. The public nature of the shrine is very important. The processes of cleansing and healing are accomplished in part by making known to the world not only what happened at this place, but also by insisting that what happened matters.

As we drove away, Michael was reminded of W. H. Auden's poem "Musée des Beaux Arts" in which a ploughman glances up at Icarus as he falls from the sky. Uninterested, the ploughman returns to his work. Icarus's failure to reach the sun is unimportant to him. It does not touch his life in any way.

At the Ormeau Road, and elsewhere in Northern Ireland where sectarian-based hatred has resulted in shootings and killings, people say with the shrines rather the opposite. They say that the failure experienced here—a failure of religion, of civilization, of humanity—is not unimportant, and we must all stop our routines for a moment and bear witness to it.

I was surprised that, given the circumstances of these deaths, the shrine contained no political or sectarian sloganeering. Since I had seen innumerable wall paintings touting either side of the paramilitary divide, I would have thought that a family member might use the public-display opportunity of a spontaneous shrine to express anger at the paramilitary group that perpetrated the killings. Michael told me then, and I have since corroborated this on several other occasions, that the people of Northern Ireland do not view the shrine as an appropriate medium for the expression of such sentiments. Michael was surprised that I asked him about this. Clearly, the shrines are an opportunity to express grief and love, personal relationship and personal loss. As we have seen, people use the shrines to address God and humanity, crying for peace and articulating only love, pain, and forgiveness. They express anguish and confusion over the situation they find themselves in, the apparently endless violence of their homeland. But they do not curse the enemy.

There is a clear distinction here, then, between these two genres of public display, the wall paintings and the shrines, and I think they should be viewed in conjunction with one another (along with the other objects and events we have examined in this book). Despite the internal divergencies within each form, and despite the multiplicity of viewerships and constructed interpretations, these two genres, as media, carry complementary and at times contradictory meanings. Certainly the gable-end murals found in Belfast and Derry and elsewhere in Northern Ireland are dramatic, striking, and highly visible. They represent important sentiments and attitudes that have to be acknowledged and accounted for. By their very nature, however, they tend to simplify a very complicated situation. One man commented, "They make a complicated thing simple for outsiders. They're not *it*. They are seized upon by photographers and cameramen, but they're too easy. They're not it." So what, I wondered, is "it"? What is the "it" of Northern Ireland? Can this place, where identity is so frequently contested and means so many different things to so many people, be essentialized?

## The "It" of Northern Ireland

The public components of ritual, festival, celebration, political demonstration, and public display are competing visualities, multiple sites of narrative struggle expressed symbolically. To really appreciate the conflict in Northern Ireland we need a cultural overview of the sites of struggle there that (sometimes fortuitously) contribute to political meaning and contribute to these narratives. We need to include the various constituencies and the various productions.

Note that the spontaneous shrines mark the sites of untimely deaths. That is, the place where the death or deaths took place—or as close to it as one can get, or a reasonable facsimile thereof in cases where the actual location is entirely inaccessible—is ritualized, as opposed to (or, usually, in addition to) the ritualization of the place where the body is interred or, if cremated, where the ashes are kept. This is an interesting development. Philip Robinson has suggested that it has to do with the decline of belief in the doctrine of resurrection of the body in Christian popular thought, thus making the place where life was lost—rather than where the body rests— the more important site of commemoration (personal communication). In the Christian nation of Ireland, both north and south, he may well be correct. Still, in marking violent and untimely deaths, there seems to be a parallel with ghost belief, which often describes ghosts as people who met with untimely deaths, who died before their appointed times, or whose death rituals were improperly administered. The ghost is the poor soul seeking adjustment into the otherworld. While there is no reported conjunction of spirits with the spontaneous shrines, there does seem to be a need to somehow consecrate the place of death in an attempt to reestablish some kind of spiritual balance that was upset through a sudden, violent, or early death. These are "bad deaths," as opposed to the "good deaths" anticipated and prepared for with the medieval *ars morendi* rituals (see for example Muir 1997:45–48). In her study of items left at the Vietnam Wall, Kristen Hass talks of the memorializing work of leaving personal items as mediating the dead and the living in an attempt to create an appropriate memory; that is, to make sense of a senseless war, and to make meaningful in this context the ultimate sacrifice (Hass 1998). The situation in Northern Ireland is very different, but the victims are always viewed as undeserving of their fate by those who leave their offerings (even when the deceased are military personnel or paramilitaries). So in both cases the spontaneous shrines mediate, or appease, the unquiet dead, and help put the living at ease.

Eileen McManus, whose father was killed at Sean Graham's, was the first person to post a public message on the building in which he was killed. Her simple message, "Daddy, I have always loved you and will never stop loving you," was so poignant that a photograph of it was carried in major British newspapers. Eileen McManus was 14 years old at the time, the youngest daughter in her family. Her father, Willie McManus, was 54. Eileen's actions sparked others to write messages to their lost loved ones as well.

**Figure 19.** "Daddy I have always loved you": Eileen McManus's Note to Her Father

Eileen's mother, Mrs. Roseleen McManus described to me her daughter's role in the creation of the spontaneous shrine at Sean Graham's:

> R.M.: She started all that off. Did you ever see the flowers? That started off there up the bookie's and then everybody that was killed, they put flowers. She tied a wee rose round his—did you see it there? She tied a rose, then they all started doing it, you know? It was just something she felt she had to do.
> J.S.: Did you put flowers down too?
> R.M.: I didn't go near it. I still can't go near it. I can't go up—if I go out I go up the long way. I walk around it. It's like a monument to me. I'll go up past it but I'll look the other way. I've never looked at it.
>
> I think she wanted people to know if anybody read it, seeing what the people feel, and seeing what she felt for her father. You know she loved him. She never thought that was going to happen. Maybe the ones that done it, you know, maybe they'll soften. But they'll never soften. It's just, that's it, just write a note to them.
>
> Eileen put the rose up, then they started bringing wee flowers and stuck them on the ground. Then all the flowers came, and then two days— Wednesday, Thursday, on Friday morning—the flowers came from Dublin. Brought all the flowers up. All Dublin. Did you see the flowers outside? They

all came from Dublin. Flowers are something if you want to show your feelings you always give flowers, you bring your wife a bunch of flowers.

Eileen herself told me, "I wanted to show people what they had done. That's it. It was mostly for people around here, people who already knew him. [I] put flowers there at Christmas and the anniversary. Just something I feel I need to do for myself."

Later, Mrs. McManus told me:

> She's only getting out of herself these couple of months back. That's why I like her to go out to aerobics... She went away to France for six weeks. It helped her a lot, going to France. Another thing—we used to go to England. I used to go to my family every year. She used to write him notes every day. I think maybe that was the back of it. She'd write him wee notes saying she was keeping all right, and funny enough for the two of them. I got a note there in the bag she wrote to him. She was always writing a note to him when she was away. Maybe that's what made her do it, you know, saying, "Well, he's gone for the last time." She wanted to tell him she loved him.
>
> He wasn't working. He used to be a builder. He was 54 when he was killed. She was always with him. He nursed her, and fed her, and changed her, took her out, to school. The bond was there between him and her, the father-daughter bond.

Interestingly, Eileen and her father frequently wrote notes to each other while he was alive, and thus Eileen's final note was an extension of a private family tradition. We have here a personal action motivated by personal needs and traditions, but one that was also a social, communal act which generated similar notes from others. Eventually, Sean Graham's attracted symbolic reactions in the form of Orange parades and demonstrations. In the summer of 1995, after the ceasefire agreements of the IRA and other paramilitary groups, the Orange Order was denied a permit to parade down the Ormeau Road. It is important to note in the testimony below that Mrs. McManus talks of the parades as being in the same conceptual universe as shootings and bombings. She recognizes them as invasive, offensive actions. The power of these events, which may appear harmless to an outsider, is not lost either on her or on the organizers and participants. Also, I cannot help but notice that she refers to her husband in the present tense.

> [After the ceasefire of 1995] there's not going to be no shooting or no bombing but there's still going to be trouble. I mean they can't agree with each other, you know. They've no right to come down. It was the mini-Twelfth [July 1]. Do you remember the first year they gave the five fingers and all that? It was very bad here when the Orangemen come down. They lost the

right as a result of they danced and all as they come by the bookie's, and people died there. It wasn't five animals. It was five human beings.

So they thought maybe if they would stop that particular one. But they're all the same. Why let an Orange parade march through [an] all Catholic area? It's like a republican band coming down the Shankill Road. It wouldn't be allowed. So if we're going to get equal rights here they'll just have to all stop. ALL the parades must stop. They've had their Orange parades for hundreds of years. It's time to call it a day.

The police suggested they take another route. They'll never come back down here. And they won't like us for that. In saying that, I've never been to a protest. I've never been to a protest. They've always come down, and my husband won't allow me to go near it. Let them go down. And now I've never been, but I don't think they have the right to come down. They've lost the right. I mean they never sent—The Reverend Martin Smith, he's our MP [Member of Parliament] for the area. He never came to say "Sorry" to any of the families. He just completely ignored us. He never came near us. Nobody came near. Now the churches, the wee churches were very good to us. The wee church up University Street and a load of churches, ministers and all came up to the house. But the Reverend Martin Smith, he's over the Orange Order, he never showed no—and he seen what they done.

## Gesture as Political Ritual

The "five fingers" that Mrs. McManus refers to is a particularly offensive gesture: to raise five fingers while parading in front of the murder scene is to say "Protestants 5, Catholics 0." This keeping score, so to speak, is very common in Northern Ireland. For instance, the Ulster Protestants claim the ancestry of something like 25 American presidents. When asked how he felt about John F. Kennedy, one man told me he felt like, oh well, Protestants (or "Prods") 25, Catholics 1. When a man in the parade held up his five fingers, he was widely and correctly interpreted as saying, in essence, "We got five of yours." Mrs. McManus rightly understood the insensitivity and viciousness of the gesture.

We see here the power of gesture. Gestural politics is of great importance throughout Ireland and in Northern Ireland. Smaller rituals such as shaking hands take their place with larger ones, such as parading, within the entire spectrum of public symbol as political communication. This is an important area that I will address at some length. Along with the strong sense of historical grievance that imbues residents of Ulster, gesture as ritual, as stylized statement, is a dynamic factor in the ongoing aggressions. For instance, when Sinn Fein leader Gerry Adams shook hands with British Prime Minister Tony Blair on October 13, 1997, at the beginning of formal peace talks, it was international headline news. On the Cable News Network

(CNN) it was reported that they shook hands at Stormont, in Belfast, which was the home of Ulster's previous locally based parliament and would again serve in that capacity as a result of the negotiations. However, CNN had no footage of the actual handshake. It took place away from the cameras, in private; it was intended neither for the media nor for public viewing.

Even as commonplace a ritual as shaking hands is enormously important in this political context. Some months earlier, on a trip to the United States, Gerry Adams debated Ulster unionist Ken McGuinness on CNN's Larry King Show. Adams offered his hand to McGuinness, almost challenging him to shake it, supposedly as a sign of good faith. McGuinness refused, and in fact tried throughout the broadcast to avoid looking directly at or making eye contact with Adams. I have noticed this on other occasions as well. It is as if his rivals, by refusing to acknowledge his presence, can deny his legitimacy. Adams is widely understood to be the leader of the IRA. Most unionists (and many nationalists) consider him a wanton murderer. As described by one very moderate individual, a Protestant but one not opposed to reunification with the South, Gerry Adams knowingly sends 18-year-olds to their deaths and then helps carry their coffins in a hypocritical show of solidarity. Because they view the IRA's activities as illegitimate, murderous attacks on their own people, unionists refuse to acknowledge the legitimacy of Adams as a political leader. And because of their deep sense of historical grievance, they refuse to even shake his hand. Adams is aware, when he extends his hand to McGuinness on American television, that he puts McGuinness in an untenable situation: Either he forsakes his deeply held principles or he looks bad on American and international television.

In an interesting way, this refusal to participate in a social ritual as a way of denying another's legitimacy is paralleled by Adams's refusal to attend sessions of Parliament when he was elected MP from West Belfast. His point was that his empty seat was a sign to all that, according to the views of those who elected him, the British government had no authority or dominion over the six counties of Ulster.

In the United States, handshakes are freely given and no longer carry the burden of contractual assent ("We shook on it"). Handshakes between even antagonistic heads of state routinely provide photo opportunities for journalists and media consumption. But in Northern Ireland, handshakes are a matter of great principle. Thus when Blair meets Adams in Belfast to begin peace talk negotiations, the fact that these two leaders of what are, in effect, warring states shook hands is in itself highly newsworthy. Significantly, though, the image of the handshake is potentially so problematic for Blair in a political sense, and so morally repugnant to unionists, that it was hidden from view. In fact, Blair was heckled and threatened as he walked through a shopping mall in East Belfast later that same day. The fact that no pictures were allowed of the handshake speaks to the larger point of this book: that visual ritual display is of crucial importance in the public discourse having to do with issues of the most profound importance in Northern Ireland.

In addition, Mrs. McManus objected to the absence of a personal visit or any display of condolence from her MP, a well-known leader of the Orange order. Whatever the reason for his lack of a visit, it was deeply offensive to Mrs. McManus. Otherwise she is careful to point out that Protestant churches formally noted her loss, and Protestant friends did also:

> The local church sent me a beautiful bouquet of flowers and groceries, you know, to help us out, there was people coming. You couldn't have moved in this house from Wednesday until he was buried. He had a lot of friends. They sent us soup and all. Showing that they were being kind. It's very good. All Protestant people are not bitter, you know. Just like both sides, there's bad on both sides. I've got a wee friend I used to go to bingo with and, now, she came from the Lisburn Road. But I just got to know her through going to the bingo and we were calling on Thursday night when three people were shot on the Tuesday before the bookie's, and she and I were condemning it. She was condemning. She was saying it's not fair killing innocent people, and she was a Protestant. I happened to meet her the next night and all the girls who work in the bingo office sent me a sympathy card saying "We don't approve of what happened." No, it's not all one-sided.

For his part, Mr. George Patton told me that in fact the Orange Order had offered to lay flowers at the site of the slayings as they walked passed Sean Graham's, but the offer was refused. He also insists that any objectionable behavior during parades by any participants is not approved by the Orange Order and is dealt with swiftly and sternly. Nevertheless, he fully believes in the right of Orangemen to parade on the Ormeau Road, and anywhere else in Ulster. In this case, a significant gesture would have been giving up the parade route entirely, something the Orange Order and unionists generally are unwilling to do. Laying flowers at the site, while appropriate for many, is not enough as far as the friends and family of the victims are concerned.

> J.S.: Would it be true that if they were parading past a Catholic church they would make a bigger racket, that sort of thing?
> G.P.: The rule of thumb is that if there is a service on, you don't play at all. We teach our people to be respectful. If the building is empty, then it is treated just as an empty Protestant church and you just blather away, as they say. But I don't think they would increase the tempo. I don't think they could, to be honest with you, even if they wanted to. We've had problems in the past, not with Lambeg drums but with certain bands who would stop to play outside a Roman Catholic church and would dance and would play the bass drum a lot louder and things like that. We deal with that because we don't want that to happen.

J.S.: So officially you would try to stop that sort of thing?

G.P.: Oh yes.

J.S.: But it does occur on occasion.

G.P.: Has occurred. And if it does we probably make sure the band doesn't walk again.

J.S.: Really. Would you just give them a warning the first time?

G.P.: It depends how seriously they've misbehaved. If it is appropriate to warn them, we warn them. If it is more than that we say, "Right boys, thank you, good night, you'll never walk again with us." And that happens. We're very conscious of that. We believe in the right to parade, intrinsically believe in that, but we also accept that with the right comes responsibility. So if anyone is messing about in our parades we make sure they don't parade with us again.

J.S.: Do you ever stop and deal with the basic fact that, no matter what you do, there is going to be a negative perception, or by the fact of doing it you're going to hurt people's feelings?

G.P.: There are those people, no matter what we do, [who] will take offense. People travel miles to be offended. What people often forget is that Northern Ireland is a relatively small area and we know each other. For example, there was a recent parade a few years ago [which] created a lot of problems. And we haven't walked down the Ormeau Road with that particular parade ever since.

J.S.: Is this the Sean Graham's thing?

G.P.: Yeah. We condemned the murder at the time. It was a terrible thing to do, as we condemn all murders. We've condemned it, we've said those people don't speak for Protestants and so forth, they have no place in our community. But it has nothing to do with us. The Orange Order is not involved in murder and stuff. This is an old traditional parade. So we parade and we parade peacefully. Unfortunately there were a few people who didn't parade peacefully and that created problems. We dealt with those people. One of the big problems we have is that we can only deal with people after they've done something. And sometimes that's too late. The damage is done in people's perceptions. But on that occasion there was a crowd there, protesting the parade. And we knew that a number of those people had come from a very, very long way to protest. They weren't the local concerned citizens they claimed to be. So there is nothing that we'll ever be able to do to convince these people. What we have to do, and what I hope we are trying to do, is to convince, or to help people to a better understanding, including our own members, our own people, of what our parades are all about. It's a responsibility of ours to say, "Look, this is our culture, this is our heritage, we don't mean to offend anyone, we certainly will do all we can not to." In terms of our parades, it's a point very often forgotten or neglected, we actually only parade where we are invited. There is always an Orange presence in the area, and the local Orangemen invite the others to come. We never parade through an area where there are no Orangemen. We have no intention of parading across the Falls. There would be no reason for doing it. That would be offensive to people, that would be triumphalist. But we do believe that where there are people in our community they have a right to express their culture.

J.S.: Did you consider voluntarily rerouting the parade as a sign of respect?

G.P.: The parade was going to lay a wreath on the bookie shop as a mark of respect. But we were then taken to court, and it was said, "We don't want you, end of story." Now, unfortunately in Northern Ireland, that gets people's backs up because the one group of people the Ormeau Road residents did not approach about the parade were the organizers of the parade. They wrote to 150 community groups, they wrote to the police, they took us to court, but they never once went to the Orange Order and said, "What are the plans?" And we had plans. They just went off on a tangent. And our people said, well, if that is the sort of people we're dealing with we probably cannot stop the parade. It's unfortunate. Very unfortunate. But that's Northern Ireland society, unfortunately. We're not white on white. We do try, I have to say we do try. I accept there are those in the Roman Catholic community who would try to make similar gestures and are prevented because of the attitude of people from my background. That's very regrettable, it's terrible. The one important thing to come out of it: Don't stop trying to make those gestures. Let's keep going.

The amazing thing, I think, is that we recognize that in both Roman Catholic and Protestant communities the terrorists are a small voice, but they are a very vitriolic voice. I can appreciate how difficult it is for people. You know these guys with the guns, you know they don't think twice about knee-capping you or shooting you, even if they are from the same community as you. So it can be difficult to publicly say anything. But when you get the opportunity, God forbid, I mean I hope there'll never be another opportunity to lay a flower or something at anyone's—the spot where anyone's been mur-dered. Your true feelings are then poured out that you find it may be difficult, in normal circumstances, to express because these guys are going to come down on you.

I happen to believe the provos [Provisional IRA] or the UVF do not simply inflict hurt on the opposing community—as they see it, the opposing com-munity. They inflict in some ways possibly more harm and more damage on their own community. They're gangsters and thugs, that's the bottom line. And they inflict terrible damage on their own people, through extortion, through blackmail, through punishment shootings, through the lack of investment in areas. People aren't going to invest in areas where their factory is going to get blown up or what have you, so they are suppressing their own people. But it's very hard for an ordinary person to stand up against some thug in a mask who is carrying a baseball bat or a hurley stick, or even worse a Klasnekelov [automatic weapon], and say "I don't like what you're doing." You can do it in other ways. And that's what—this is my opinion, and that's all we can express—I think times like this the true feelings of the people are expressed. We're not saying "We hate the Prods," or "We hate the Taigs," we're saying "Let there be peace!" The flower shrines, yes, it's the voice of the people, which can find difficulty in expression in any other way whilst you have these hoods.

## The Voices of the People

After the infamous incident at Sean Graham's, the peace talks almost collapsed entirely when the Orange Order was stopped from parading in Drumcree in 1996. Rioting followed. The Royal Ulster constabulary reversed its decision, which then led to rioting on the part of nationalists whose neighborhoods would be transversed by the parade. This is an example of the power of public-display events, festive events generally and parades specifically, when used as weapons. As Michel de Certeau suggests, parades are invasive (de Certeau 1984). In Northern Ireland especially, the rites of intensification of one political group are necessarily seen as conflictual and offensive by the other.

The shrines are different from the parades in important ways. People agree that there is an inviolable sacredness to them. For instance, Rhonda Paisley, who curated an exhibition for the Ulster Folk and Transport Museum on symbols in Northern Ireland, and who is the daughter of the Reverend Ian Paisley, had this to say:

> I think that everybody thinks that you don't pass something like that without thinking that somebody has taken the time, and cares enough, and you know I think there is a sacredness about it as well, that people have respect for life and death. I did relief work in Romania at the time of the revolution and the shrines were literally this high. I mean there were heaps of them piled high at lampposts and trees. And the little pictures, just like passport-size pictures, pinned up where somebody thought that was probably the spot that their relative was at when they were dying. And there were little candles burning.
>
> I also feel it's done [in Northern Ireland] by people who don't feel as if they could go to the person's grave because they're not part of the family or they never knew the person. But they can pay tribute in their own way at where this happened, because it's a very public place. And I think you see mothers taking children with a bunch of flowers to these sites, and you see that their children are being taught, you know, we're honoring the fact that this is a person's life having been ended so brutally. And they can't maybe take their children to the funeral procession of six victims who were bombed to death. But they can take them to where the bomb happened and let them show respect there. You know, I think there's sort of, in our country anyway, whether that's true elsewhere, but I think there must be a sense that, you know, I can go there and do that as my token, and maybe I don't know the person and can't go to their home or won't be attending their funeral, but this is what I can do.

Along with simply not knowing a victim personally, differences in religion might make home or church visits uncomfortable. In addition, Ms. Paisley refers to taking children to the shrines as a means of teaching them respect for life. Socialization occurs at the shrines, just as at the parades and bonfires, but it is a very different kind of socialization.

So far we have seen at least four emic, or insider, theories on the reasons behind and purposes of the shrines: (1) Both Eileen McManus and her mother suggested the purpose was to communicate to the people who perpetrated the killings, and also (2) to communicate with the deceased victims; (3) implicit in this is the intention to communicate one's sense of grief and loss to people who knew the victim(s) and who live in their neighborhood; and as expressed above, (4) the idea that the shrines allow people to anonymously express forbidden or unsanctioned sentiments. I have seen bouquets of flowers left at sites where, for instance, Catholics were killed, with notes that read, "From a Protestant family." The shrines allow those who disapprove of the violence employed by the paramilitaries to express their feelings with less risk of retaliation or intimidation by members of paramilitary groups. More than

Figure 20.   Spontaneous Shrine for Protestant Victims

one person has suggested this to me; and in a country where, even after an official cease-fire, punishment beatings and shootings continue, this interpretation must be taken quite seriously. These suggested motivations for creating shrines are complementary rather than competing functions: they can and do coexist simultaneously within and among those who create memorials.

I worked closely with a young woman whose husband was murdered by the IRA only a couple of months before I met her in August 1991. In the following testimony, Mrs. Lorraine Lawrence discusses the spontaneous shrine for her husband, and her reasons and motivations for maintaining it. They closely parallel those of Eileen McManus. Keep in mind that Mr. McManus was killed by loyalist paramilitaries whereas Mr. Lawrence was killed by republican paramilitaries. Nevertheless, the shrines are similar in appearance, purpose, and sentiment:

L.L.: His murder? He was shot. It was three years on June 17. He was going into his work. Just as he was parking his car. It was just outside the docks area, it was a place called LTS. It was a very built-up industrial area. And we call it the docks area really. Lot of comings and goings that time of the morning. So I think that is why it was easy for them to risk it.

We were married almost eleven years. Sorry, no, almost ten, that year. It would have been ten on October 12th.

J.S.: Do you have any idea why he was killed?

L.L.: I'm absolutely certain why. It was [his] part-time UDR work. Ulster Defense Regiment. It's now the Royal Irish Regiment. It's come under a new title, so if you hear it on the news you'll know. Well, it's just the police and the regular army.

J.S.: It's both police and regular?

L.L.: Uh huh. That's their system, really, especially the part-time force. I think they still have the part-time force. There was talk of them abolishing it, but I think they still have it. I'm not sure. Not in touch with too many people now.

J.S.: Would he have had to serve six months every three years or something like that? In the United States there is something called the National Guard. And you have to go on what they call active duty every so often for a short period of time.

L.L.: Yes—No, No. He did a camp once a year and that was just a week. And we have to take a week's leave of work for that. That was part of his holidays. But he didn't mind. He enjoyed it. Something he enjoyed so much, it was a part of his life, you know. No, that was the only thing. They had to do their initial training. And then they just did, they got pieces [of their training] at weekends and different training sessions and that. It wouldn't have been as tough probably as the regular army.

J.S.: Is the idea that they could call them up in an emergency.

L.L.: Oh, it would have to be an emergency. Or ... if the country went into civil war or something like that, they would be called up. But only in a dire emergency.

J.S.: This is obviously sensitive now. Had there been any previous threats or anything like that?

L.L.: There'd been no previous threats, but it was definitely the IRA. I think they did claim responsibility.

J.S.: Do you think this was a personal thing. Or do you think it was just two people you never met in your life? I say two because it's usually two.

L.L.: Oh, two people. I think there was three involved all together. It was definitely three people who we've never met. Obviously somebody had pinpointed Brian. The three must have been following him for some time to find out what route he took to work. There were only two routes you could have taken and we're all creatures of habit. And I think he parked at more or less the same spot everyday, and they knew. He actually opened up—I wasn't even aware of that at the time. He opened it up in the mornings and I wasn't even aware of that, I thought the manager did that, but apparently not. And it was because of that reason. Oh, definitely. You know, I mean I had heard of the terrible things that were happening that year when there were taxi drivers being murdered. Particularly that year. I mean it's still going on of course. June '91.

I suggested for him to leave it, but he wouldn't. It become so much a part of his life. And it was something he really enjoyed. He was reluctant to leave it. You know. I had this sense of things were getting really awful and you know this could happen, but I didn't think it would. But I mean I am not psychic or anything. But I asked him. I hadn't thought about it when he joined in 1982. I mean you didn't think. You thought this was something that would bring in extra money. And we weren't that long married ... He wanted to join the Royal Air Force, and he had an asthmatic problem and then of course the qualifications you need for that even back then. They required a lot. So this was sort of second best, you know, to join that. You know, when he was younger he wanted to join that.

J.S.: Was he political at all?

L.L.: He wasn't, no, no he wasn't. He would have mixed with both sides of the community and, you know, he came from Holywood, close to here. You know, he said some of his best friends were Roman Catholics when he was younger.

I'm from Belfast. Originally from North Belfast. Now I'm living in the east [Belfast]. When we got married and were living in the same house ... we moved to the east side of the city ... Ravenhill Road. I'm originally from the North Crumlin Road.

J.S.: Would you feel the same way? That you never had any real sense of bias one way or another?

L.L.: No I haven't.

J.S.: Does this event change that in any way?

L.L.: Well, it did a bit at first. I mean you have to live here. Then not so much. It would be more directly against the people who did it.

J.S.: You mean the IRA?

L.L.: I mean the people who did it, and the people who wanted them to do it. It would be more them. Obviously them under the organization than the other people who have committed similar crimes. Yes, it would be them. Of course I'm bitter towards them. But not the Roman Catholic going about his everyday life. Of course there are a lot of sympathizers on both sides, which you wouldn't agree with, you know.

J.S.: Do you feel the sympathy in a sense helps create the conditions?

L.L.: Oh yes, definitely! Very much so, yes.

J.S.: So, you know what I am talking about with the idea of marking the spot. Did that happen in your situation?

L.L.: Yes, it did. There was a wreath put down by his colleagues and then was later placed on the grave. And we went and we put flowers down and we hung flowers to the lamp, the post outside. And a couple of people from firms nearby who knew Brian. It was actually a couple of the guys [who] helped resuscitate him but couldn't. And they put flowers out front. We still do it every year. They continued to do it the first year, his colleagues. And after that they didn't bother.

My parents, just, and his mother, his mother's alive. She doesn't now. She just goes to the grave. But we do it every year still.

J.S.: Do you just put a wreath on every year, or do you put other flowers as well?

L.L.: No, we just tie flowers to the post. Just do that.

J.S.: Why?

L.L.: Just to let people see what happened in that particular spot. So they know that something tragic has happened and that will trigger their minds to maybe—They might know exactly what happened and might just drive on past and forget about it.

J.S.: Do you think the wreath is still there. It's now August. You put it on in June.

L.L.: The flowers. I don't think so. They died.

J.S.: How would you tie them? Just a string around the pole or something?

L.L.: Oh yes. Just string cord.

J.S.: Did you ever put any kind of note or a card or other things besides flowers there?

L.L.: I did put a note. It wasn't this year. It was last year.

J.S.: It was a note to him?

L.L.: Yeah. But then I thought it was a bit—It wasn't the right thing to do . . . I have written and done the odd thing on the grave like that. Then I thought afterwards . . . So I would be putting whatever I put on it without a note of any description. We'd put up the crosses in remembrance. Those. And then we would put something—and his sister—we'd write something on that sometimes.

J.S.: You'd pin a note to it?

L.L.: Well, you can write on them actually, just in pen. Onto the crosses.

J.S.: Are they flower crosses? Am I thinking of the same thing?

L.L.: They're just a wee plain cross, really. A remembrance cross.

J.S.: What would be the difference in your mind between doing something at the grave and doing something at the spot? Why do you do both places?

L.L.: Probably the grave is more personal than the other place [which] is more public as you say. There wouldn't be many people unless they were coming to visit relatives' graves. And they would just pass it by and notice it. But apart from that where you've got a very built-up industrial area, as I mentioned earlier, you got everybody going about their everyday business.

J.S.: So once again, you warned everybody going about their everyday business to at least be aware that something happened here. Do you know other people who are in similar circumstances? Have you, for instance, joined any organizations for people who are victims of violence?

L.L.: I haven't, no. I went to a therapist for a while. Privately, you know. He was a psychotherapist. I had him privately, but I haven't joined anything. Maybe I should have but I haven't. No, I haven't.

J.S.: Do you have any general observations as to why people do that? Because it is pretty widespread. Not the killing, but the marking the spot with flowers. Both at the time of death and a kind of annual commemoration. What do you think spirits the impulse initially when someone is killed on the street?

L.L.: We didn't do it right away, now. It was the day after the funeral. We did the funeral Wednesday. He was killed the Monday morning. The funeral was Wednesday afternoon, and then it was Thursday before we did it. I haven't actually thought an awful lot about it, but it was my mother. She wanted to do it. It was really her, I sort of went along with it.

J.S.: Did you say the wreath came from the funeral or vice versa?

L.L.: The wreath was laid there first of all and then it was eventually placed on the grave by his colleagues. Then they took it on down. Most of them were, of course, at the funeral. It is to mark what happened: A tragic event took place on that spot.

J.S.: I was struck by [the fact that], even though there were notes and so forth, there was no what I call "political sloganeering." Why do you suppose people never express those kinds of sentiments?

L.L.: Probably because it would be putting themselves down. Not really, I don't know. Not really. It would just be the wrong thing to do at that particular time. You're marking tragic events, tragic loss of life. It shouldn't have happened, and this is really to keep the two separate—you know, the loss of your loved one and your own feelings about who did it, who's responsible for it. Wouldn't really be the right place.

*J.S.:* What would be the right place?

L.L.: The media probably. Go talk to them and try to arrange some type of interview. Not that it does any good. People have done it.

J.S.: Would you be any way offended if you were walking down the street and you noticed the curbstones were green, white, and gold?

L.L.: On my own street, oh yes, very much so [laugh]. It's a different religion. And I would feel very frightened probably. Intimidated by it.

J.S.: But if you were walking in another neighborhood and you weren't thinking much, but it turned out to be a republican neighborhood or just a nationalist—I guess there's a slight difference in the terms—would you feel funny about that? If you noticed the curbstones were painted?

L.L.: If I wasn't expecting it, yes, certainly yes. I would feel uncomfortable. But certainly if I was expecting it, it wouldn't bother me. I probably wouldn't be there [laugh] to start with. I would avoid it.

J.S.: Is that more so since the unfortunate events?

L.L.: No, No. Just taking care in Northern Ireland.

J.S.: Would you feel similarly in a neighborhood that had red, white, and blue curbstones?

L.L.: Oh no. Same religion, yes.

J.S.: What are you—you're Protestant, but are you Presbyterian?

L.L.: I don't really belong to any church.

J.S.: But you are Protestant in the general sense. If you saw a house and it had the Irish national flag flying, would you automatically say that there were IRA sympathizers in there? How would you read it?

L.L.: Yes, probably. Normally in a Roman Catholic, you know, in a middle-class residential area, they wouldn't put a flag of that description up. So I would read it [as]—if they weren't involved in an organization—certainly sympathizers. Supporters, definitely.

J.S.: Then you would probably have negative feelings, given your experience.

L.L.: Well, we're still members of the United Kingdom, so we should be able to fly the Union Jack. And the south of Ireland should be able to fly the Tricolour, you know, because they're not members of the United Kingdom. We should be able to, you know.

J.S.: Do you think there will come a time when you will not commemorate annually the event? Let me ask you a little bit more about that. Do you make any kind of little ritual—I mean, would you say a prayer? Do you just go down there at some point and tie flowers and leave? When do you actually go about doing that?

L.L.: We just go and tie them. I'm not religious.

J.S.: So the flowers have no religious significance for you?

L.L.: They may do to my mother.

J.S.: But you still find it important to do it?

L.L.: Oh, yes.

J.S.: Do you just do it at a convenient time of day? Or do you do it a certain time of day?

L.L.: Just whenever is convenient.

J.S.: Then do you just sort of stand there for a few minutes in silence?

L.L.: Just a few minutes, yes.

J.S.: Now, you did it this past year?

L.L.: Yes, we did it.

J.S.: Would you tend to assume you will be doing it again next year?

L.L.: Oh yes.

J.S.: At any other time during the year would you visit that spot? At Christmas?

L.L.: No, not that spot.

J.S.: And would you expect to carry on doing that for the indefinite future?

L.L.: I would expect so, yes. Unless I left the country or something like that. That would be the only time.

J.S.: Does it make you feel better to do it?

L.L.: Oh yes, it does.

J.S.: Does it bring back the memories?

L.L.: A bit, yes, but because I wasn't actually there I didn't actually witness it.

## Folk Memorialization and the Personalization of War

Here and above I have viewed the parades and the shrines, and the murals as well, as competing public symbolic forms: dialogic symbolic statements responding to other artifacts in the same category (the shrines reference previous shrines, the murals other murals), and across genres/media/categories as well; the shrines can be read as responding to the murals and vice versa. Murals frame death, for instance, as patriotic, national, and heroic; shrines frame death as personal, familial, and tragic. Although both perspectives are true depending on who you are and where you are standing, they are very different political statements. If many murals are essentially recruiting posters for the various paramilitary groups, then the shrines answer these by demonstrating the results of that paramilitary activity in personal terms.

Spontaneous shrines are a kind of folk memorialization, temporary in nature but capable of being renewed, revived, or reanimated on meaningful dates and occasions. Recent work on official war memorials has tended to view them as shrines designed to be permanent edifices. There has been a notable movement toward increased individualization in civic memorials to the war dead, away from memorializing "those of name" and toward naming all the soldiers themselves (see Laqueur 1994). This movement toward personal identification has led to sites such as the Vietnam Memorial in Washington, D.C., on which is inscribed the name of every American individual who is known to have died as a result of fighting in that war. Additionally, however, this official memorial has become personalized—it too is the site of folk memorialization. People leave notes and personal memorabilia of their loved ones at the site. Whereas the Vietnam Wall is neither the place where the people named were killed, nor where they are buried, it does function as a quasi-tombstone. Indeed, it looks like a tombstone, a slab of granite bearing the names of deceased people on its surface. It has become a kind of spontaneous shrine itself, and it bridges the distances between official memorials, graves, and spontaneous shrines. As to a grave, people bring flowers; as to a spontaneous shrine, people bring photographs and personal memorabilia. The Northern Irish spontaneous shrines memorialize nonmilitary victims of a nonwar, an armed civil, guerrilla strife. They are temporary, not permanent; erected by families and friends, not municipal committees. Those memorialized are more often seen as innocent victims rather than martyrs. However, both the Vietnam Wall and Northern Ireland's spontaneous shrines represent the personalization of war. They also provide sites for the popular critique of war.

The relationships between personal, private acts and public, communal ones are evident here, especially in Eileen McManus's actions. These forms have powerful social meaning, but they are created and sustained and understood by individuals. The memorial shrines are the result of the activities of many, but the shape of this particular one was heavily influenced by a 14-year-old girl. Since symbolic forms are multivocal and polysemous, individual interpretations and perceived meanings vary (Turner 1967). Yet, since they are derived from a shared symbolic vocabulary, the range of interpretations is generally restricted (Charsley 1987). The study of this family's particular and specific use of traditional symbolic forms to respond to trauma and to express shock and grief must be conducted within the larger context of culture and politics in Northern Ireland (Glassie 1982). An interpretive ethnographic study must be based on an understanding of local knowledge and people's own interpretations of their motivations, their understandings, the ways in which they create meaning and use cultural forms. Eileen might have been motivated as much by her own personal tradition of communicating with her father in notes as she was influenced by her familiarity with other shrines. For her it was an extension of a private tradition into the public sphere; it was simultaneously a communication to the deceased himself, to those who knew him personally, to those who did not know him but chose to end his life, and to the larger public who would pass the shrine in the street. This was a personal use of social tradition, the nexus between private life and the public display of symbols; and it indicates the multiple intentions behind, interpretations of, and spectatorships for the same actions. The examination of spontaneous shrines raises issues of how people actually use art and tradition as weapons (Rolston 1991); and how a particular individual and family actively use an emerging but still traditional form, the temporary memorial shrine at the place of a violent death, to express personal grief in a public context, thus making a social statement.

The war in Northern Ireland is fought on many fronts. This one has to do, perhaps, with the struggle over popular ideology.

# CHAPTER FIVE

# Conflicts

Much to the frustration of those who live in Northern Ireland, conflict has become synonymous with the name of their country internationally. I once attended a special program with three members of the British Parliament in London, a program especially for Fulbright scholars. When I asked a question about Northern Ireland, the moderator attempted to go to another question while the three elected officials squirmed. Finally the Labour MP said, "There's nothing we can do about Northern Ireland. Those people are crazy. They're all killing each other over there. What can we do?" Certainly I find that when I mention my research interests to friends and colleagues in the United States, I am inevitably met with incredulity and comments involving bombs and danger, usually intended as humorous. In addition, residents of Northern Ireland complain that the only images the international media deem worthy of broadcast are images of violence, images that create and sustain the stereotypes of a violent people and a hopeless situation.

Again the irony: The contestive populations in the north of Ireland are viewed as a single people by the English; the Ulster Protestants are deemed Irish and subjected to the traditional stereotype of the Irish as irrational and prone to fighting. Of course, beyond the attempts of the Protestants and the unionists to demonstrate their Britishness, there is the irony that it is precisely because the populations do not believe themselves to be one people that the strife continues. That is, the population is not homogeneous, as outsiders—including the British—seem to think.

Paradoxically, it is equally common to view the situation as simplistically dualistic: Protestant versus Catholic. As I outlined above, the opposing identities are much more subtle than that. The conflict is not between two Christian denominations who simply cannot get along. The religious denominations are only one aspect of deeply conflicted national identities. The Ulster Protestants and the Irish Roman Catholics occupy the same land but claim it for different national states. "Unionist" and "nationalist" are the more correct appelations, terms that correspond with but do not absolutely imply religious affiliation. The militants—loyalists and republicans—again correspond to the religious signifiers in a crude way, but in fact these groups are usually at odds with all church doctrines and representatives; moreover, there is fierce competition and ill feeling among the militants and more moderate people and groups.

Ulster is in reality multicultural and pluralistic, although this may not be as obvious as in, for instance, the United States, where ethnic and racial differences derived from nationality are used to gloss other identity differentiations involving class, gender, sexual orientation, region, and, too, religion. This is perhaps not even obvious to the residents of Northern Ireland, but it is important to realize that there are many different groups and identities that individuals have available to them. Not all of them are as relevant to our discussion—a politically conservative Presbyterian might also be a member of Greenpeace and a feminist, for instance. And it is true that many of these identities are coded according to one's background, such as membership in different Catholic or Protestant sports and fraternal organizations.

It is also true that the violent conflict might be the most striking aspect of Northern Irish culture, or at least thought to be the most media-worthy. This is due, in part, to the very theatricality of the bombings and shootings themselves, a point very much understood and exploited by the IRA and other paramilitaries (see Feldman 1991). This is a point not to be lost in a study of the political discourse of public display. The explosions are usually intended to cause serious injury only to buildings and property: Some of the stated reasons for the IRA bombing campaign in Belfast were to disrupt business as usual, to scare away shoppers, and to demonstrate to the world that all was not well or normal in this place. Likewise, punishment shootings and beatings are meted out to enforce IRA rules concerning fraternization with British soldiers and to punish informers, but they also serve as warnings to others. These acts are certainly much more than expressive. People are killed and maimed as a result. I am not claiming them as ritual or even ritualistic, as Feldman does for shootings, arrests, house searches, and interrogations (1991), but they are theatrical public events intended to be viewed by multiple audiences.

The terms "ritual," "festival," and "celebration" also carry their own standard, perhaps stereotypical, significations. Again, when I attempt to explain my research to others, including scholars in related disciplines, the general assumption is that I am myself involved in a celebratory project; that these terms always imply socially cohesive values, and, by implication, that I am naively examining superficial and misleading events. However, celebrations frequently involve and are the vehicles of social conflict. One example is the fragility of the peace process of the late 1990s that was periodically in danger of unraveling over the issue of contested parade routes. Another example is the uproar over the way the British monarchy handled the funeral of Lady Diana after her untimely death in 1997. Here, I want to explore the clash of symbolism inherent in the public drama over the death of Diana so as to delineate this point, and to contrast it later to the clash of traditions in Northern Ireland. The latter clash occurs on the same social level. The former clash differs in both form—royal versus popular—and content, while in Northern Ireland, the forms are similar but the content differs.

### Clash of Traditions: Royal versus Popular

International reaction to the death of Lady Diana took many forms. In the United States I was surprised at the extent to which it was used as a vehicle for United States boosterism. One television commentator after another spoke of the superiority of the American system of democracy as opposed to the British monarchy. The outpouring of emotion on the part of the British was taken as a sign of a democratic chafing against monarchism. Some on-air personalities and reporters, in shocking displays of ignorance, actually went so far as to discount Great Britain as a democracy entirely. I heard at least one political talk show host say that the people of England were finally taking matters into their own hands, challenging the system, in order to accomplish what "we" had done over two hundred years, referring to the American War of Independence. The cause of this overheated rhetoric, quite specifically, was a social drama that was being played out before the eyes of the world—ironically, through the intense media coverage of the aftermath of the automobile accident in which Diana lost her life. The social drama I refer to was a clash between the popular will and the will of the Queen and the Royal Family regarding the manner of mourning Diana. It was a conflict expressed through—and constructed of—symbol and custom, flowers and flags.

That the international coverage of Diana's untimely death is ironic is because the "media," in the persons of the paparazzi, the largely freelance photojournalists who deal in gossipy, tabloid, personality-driven photographs as a livelihood, were being blamed for her death. While any death by automobile accident can be considered untimely, Diana's death took on an added dimension of tragedy as, rightly or wrongly, the photographers were demonized as villains who hounded her to extremes in the final hours of her life. Additionally, she was young, attractive, and popular in a way her former husband, the Prince of Wales, and his mother, the Queen, were not. Moreover, she was widely viewed as having been victimized in her marriage by a cold and self-centered husband. In life she easily became for many people a symbol of a contemporary, bright, active woman trapped in a loveless marriage, who chafed against a rigid Victorian hypocrisy. In death, this symbolism grew to apparently mythic, and sacred, proportions as she was constructed in the discourse of the mass media as innocent, preyed upon, the ultimate victim who paid the ultimate price (at a time when everything was looking up for her, according to virtually all accounts).

Thus her appeal to the people of Great Britain should be seen in distinction to the lack of popularity of the Royal Family. Diana seemed to be a flesh and blood individual who lived in the same world as her former subjects, albeit in a much more glamorous fashion. To an extent she bridged the worlds of the royal elite, the trendy jetsetter, and the pop star, as evidenced in her well-publicized friendships with the likes of Elton John and Sting. In fact, Elton John's presence at her funeral was a sign of her popularity and her status as a popular, rather than merely public, figure. Diana blurred the distinction between royalty and celebrity. The song Elton John performed at the funeral, "Candle in the Wind 1997," was itself a form of mass-marketed but still popular mourning; significantly, it was a rewrite of an earlier hit. Elton John has since been knighted, thus further blurring the lines between the middle class and the aristocracy, between public figure and media celebrity. Also, Diana recognized and displayed compassion for AIDS victims, homeless people, and the poor. Her popularity, especially after a death that was sudden, shocking, and apparently entirely unnecessary, should be no surprise.

I am not so much interested in deconstructing the components of that popularity, as indicated above. Nor am I arguing that Diana, herself a member of the ruling class, was in any real way "one of the people." Rather, it is her very real popularity and the perceptions of the people, their construction and understanding of Diana, that interests me (see Walter 1999; Kear and Steinberg 1999). More specifically, I am interested in the clash of royal and popular symbols of mourning that I referred to earlier in this chapter, wherein I believe the tensions between popular culture and monarchical rule were dramatized; it was through this clash that the relative lack of popular affection for this particular royal family (not necessarily the monarchy itself) was made apparent, and the people's will prevailed. Quite literally, I am examining public mourning traditions as a form of cultural-political contestation.

The outpouring of the people's grief was evidenced most strikingly, most visually, in the construction of spontaneous shrines. Consistent with their visual nature, these flower shrines for Diana were frequently shown on television in the United States. However, while the people of Great Britain expressed their feelings for Diana with flowers outside of Buckingham Palace, another drama was being played out. The reaction of Queen Elizabeth as the representative of the royal family was viewed as unnecessarily cold. The Royal Family was in Balmoral, in Scotland, during the period immediately following

Diana's death, presumably grieving as a family (both children of Charles and Diana were present in Balmoral). Because she was not present in Buckingham Palace, Queen Elizabeth did not, and would not, have the flag there flown at half-mast for Diana. It is the custom never to display the national flag at a royal palace when the reigning monarch is not in residence, so Queen Elizabeth was simply acting according to tradition and protocol. People interpreted the act differently, however, viewing it as a snub of the estranged and divorced daughter-in-law. The lack of a flag on high, inside the palace grounds, contrasted sharply with the multitude of flowers below, outside the gate. This is what I mean by the clash of traditions: the royal or elite versus the popular. Both are traditional but represent different traditions; both are symbolic but represent entirely different populations.

There were other royal gaffes: When the route for the funeral procession was announced, it was much shorter than many people had hoped for. Clearly the royal family was not thinking in terms of displaying the funeral cortege as a public—or rather popular—event designed to provide the largest number of people an opportunity to view it. In fact, the royal family initially intended to have a private funeral. Eventually, in the face of public outrage, each of these decisions was reversed. The flag was flown despite the absence of the Queen; the funeral procession route was lengthened, and arrangements were made to provide for public attendance outside the church during the funeral service.

The social drama surrounding the death of Lady Diana occurred across class and status hierarchies. We see two different traditions: official (royal) display versus popular display, making conflicting claims on space and contesting for visuality, for determining and defining public statement. As a result of the way the Royal Family handled Diana's funeral, how they manipulated the symbols of mourning, we repeatedly heard predictions that the monarchy itself might not survive. That is what I find so interesting about the Princess Di story: the ways in which the marking of death, the celebration of death, is differentiated by class; and in this particular instance, how those varying traditions confronted each other. The lack of the flag (which was entirely defensible, but for which a good case was not made)—and then later its presence—served as a reminder of the royal position, the use of symbols from the top down. The flowers, visible on the street below the grudging royal symbols, represented popular sentiment from the bottom up.

It is clear from this discussion that ritual events do not always bring about the status-dissolving condition that Victor Turner has called communitas (Turner 1967; see also Drewal 1992). As a royal figure, an international celebrity, and a public figure, Diana "belonged" to many people, many constituencies. She was indeed "England's rose," as Elton John called her. She was also a media darling, a friend of rock stars and celebrities and of the poor and helpless. Her death rocked Britain in a way similar to the way the assassination of John F. Kennedy rocked the United States. The death of a major public figure is always of wider social import than the deaths of the nonfamous: The body politic is emblemized in the death. The death is public; the mourning must be public. Unlike America in the aftermath of the Kennedy assassination, the country did not unite around the official funeral and mourning rituals. Instead, differential class and status identities were made manifest in the clash of mourning styles in the streets of London. It is not really surprising that death, perhaps the central social and cultural preoccupation of our existential era, is so important; nor is it unusual that the death of a public figure should be of such great social import and be so socially disruptive. The case of Diana, however, vividly demonstrates the ways in which class struggle, identity, power, and politics are fully embodied in public ritual, particularly that concerning the death of a person whose public identity was already contested.

In contrast, the clashing traditions in Northern Ireland are not between the elite royalty and their popular subjects, but between two broad factions of the people. The style of demonstration (a term that is used for festive public displays as well as political gatherings) is the same among Catholics and Protestants—or, more correctly, nationalists and unionists: bonfires, effigies, parades, flags, and banners are all employed. Styles of special dress, of parading with bands and banners, of organizing into fraternal associations are shared by nationalists and unionists alike (see Buckley and Anderson 1988). Many people in Northern Ireland say that both sides are exactly the same because they use the same traditional modes of expression; what they miss is that these public displays signify different kinds of resistance to different situations developed out of entirely different contexts: One group identifies with the superordinate, imperialist elite, the other with the subordinate, colonized people. The resistance from the right (unionists and loyalists) is seen as a revitalization or intensification of the existing hegemonic order; the leftist groups (nationalists and republicans) understand themselves to be oppositional and resistant to that order. The leftist nationalist ideology aims at replacing the existing structures; its proponents want "Brits Out." On the other hand, the unionists—confused and conflicted by their seeming abandonment by the government of Great Britain in their opposition to the nationalists—want the country's leaders replaced with people who will support them and their position.

The clash of symbols in England due to Diana's death was across hierarchical status lines, whereas the clash in Northern Ireland is within the same classes. The performative and symbolic genres are shared by both sides. In the case of the Queen versus the people, different forms of mourning rituals were expressed in opposition to one another; in Ulster, the same forms are used by the various groups. In Ulster, there is crucial signification in the content of the message and such formal characteristics as color, whereas in London it was the choice of genre itself that was expressive.

The parallel structures on both sides of the political divide are due to the shared vocabulary of traditional expressive genres. The Ancient Order of Hibernians is sometimes suggested as the analogue to the Orange Order, although the AOH is much less powerful and, in the words of one prominent Orangeman, tamer. Of the 3,500 annual parades, 3,000 of them are unionist and Orange. Paramilitary groups view both the Orange Order and the AOH as less relevant to their causes and needs than they once were; there is a generational divide here. The murals that depict masked gunmen are commissioned by the paramilitary groups and frowned upon or actively denounced by representatives of the fraternal organizations. At the same time, custom and tradition continue to play an important role. For instance, bonfires are lit on the evening prior to an important celebration or commemoration. In the past, bonfires were lit in Catholic areas on the night before the Feast of the Assumption on August 15, known as Lady Day in Northern Ireland. Since 1972 Sinn Fein has politicized (some would say hijacked) these fires, as the event has moved to a commemoration of August 9, the date the policy of internment (the jailing of suspected IRA members without trial) was established (see De Rosa 1998, 103). The 15th remains the date of the AOH parade, and people have pointed out to me that it is still preceded by a bonfire, but separated by a week's time. De Rosa points out that bonfires are still lit to some extent on the 14th, as before, but at the same time, many young people are unaware of the connection with Lady Day. Instead they assume the significance of the later date has to do with the deployment of British troops to Northern Ireland on August 14 and 15, 1969. Here we see the simultaneous roles of tradition as paradigm (the bonfire must precede the commemoration) and history. This is also illustrative of the tensions between the church and secular politics.

We also see, with the internment commemorations, an example of a political event that is festive. Like other festive events that may or may not be political, they employ bonfires, flags, music, processions. They may also generate a sense of communitas, as when Sinn Fein marched into center city Belfast in 1994; previously, nationalist processions were not allowed in center city, whereas the Orange parades had always been.

## Festive Politics and Political Festivals

Yet another axis by which to view public events in Northern Ireland is that of festivity and politics. Political events, no matter how serious, are often festive in nature, while festive events are political in any number of ways. Events that are ostensibly festive are often exclusionary, resistant to hegemonic rule, or become riotous; political events involve large gatherings of people, often in costume, carrying banners, playing music, and so on. If by "politics" we are referring to the conventional use of the word to mean a system of maintenance for a governing structure (such as voting in quadrennial elections for one of two major candidates for president, as in the United States), than it can be easily demonstrated that most political systems are ritualistic (voting itself can be viewed this way) and marked with festive events such as party conventions and fundraising events. If, however, we intend "political" to refer more broadly to relations of power, status, and control among peoples, then we can see that issues of politics are found in varying degrees throughout the entire range of festival and celebration. Calendar holidays and celebrations display these dual features of festivity and politics. St. Patrick's Day, Easter Monday, and the Twelfth of July are the most obvious examples in Northern Ireland, and I will return to these below. First, though, a survey of American holidays will help put this principle into an international perspective, especially if we view major holidays such as Thanksgiving and Independence Day from subaltern perspectives. We will see that the dynamics that are painfully obvious in Ulster are also operative elsewhere. Moreover, an overview of festive celebrations in the United States allows us to discern more clearly exactly what those dynamics are and how they might work.

Christmas, Thanksgiving, and even Halloween are thought by the majority culture to be at least innocent and at best positive social occasions. However, non-Christians often feel that the dominance of Christmas in virtually all realms of activity at year's end is offensive, while many Christians decry its commercialization. Thanksgiving has been challenged in many quarters for the hegemonic dominant narrative it offers: the Anglocentric origins of the United States in New England. Halloween, always a celebration that incorporates an element of danger, has come under attack from groups that are bothered by the demonic imagery of the day, while contemporary witches and adherents of a number of New Age religions dislike the holiday's negative stereotype of witches. These regular calendrical festivals parallel in kind but not in extent the problems found in any observation of the Twelfth of July in Northern Ireland: The majority culture finds the celebration innocent and traditional, with "tradition" here both a positive value and a rationale for continuing the activities (see Bryan 1998). Simultaneously, however, there are sizable numbers of the population excluded by the celebrations (or asked to celebrate through them their own subjugation, such as Native Americans); these groups insist that the dominant values are being shoved down their throats.

In that the Twelfth of July is overtly patriotic and celebrates the British government, it closely parallels the Fourth of July in the Unites States. Other American calendar holidays include Memorial Day and, celebrated to a lesser extent, Flag Day and Veterans Day (Remembrance Day in the UK). These are national political holidays and commemorations that reinforce the status quo. The assumption is that the patriotic sentiments are widely shared, and specific governments and particular politicians are symbolically blurred with the idea of the nation itself, thus effacing protest as unpatriotic. If one does not partake in the rituals, one's friends and neighbors might well wonder why not. This kind of social pressure promotes the hegemony of the already empowered, and further alienates the dispossessed. Further, it renders the concerns of the alienated, however legitimate, almost incomprehensible to those who are in the majority, or think they are.

Labor Day in the United States occupies an interesting niche here. A national celebration and recognition of organized labor, it originated as an oppositional display, the American equivalent to May 1 internationally. Organized labor continues to utilize the national holiday in this way, but its resistant qualities have been blunted by an ever increasing effort on the part of the United States government to establish it as another patriotic festival. Commercial imagery such as advertisements throughout the summer months, from Memorial Day to Labor Day, feature star-spangled red-white-and-blue motifs. Further, both Memorial Day and Labor Day have become ritual seasonal markers for the opening and closing of the summer season. How organized labor deals with all this is instructive as an example of the ways in which outgroups (and organized labor is certainly an outgroup) attempt to rescue the "original" or "authentic" meanings of celebrations. For example, at a Memorial Day ceremony I once attended, a parade ended at a war memorial in a local cemetery. Salutes were fired, a speech was given. In the address, the military officer encouraged the audience to remember the "real" meaning of Memorial Day—more than a long summer weekend, he said, it is a commemoration for those who died fighting for their country. He reminded me of a priest exhorting his congregation to remember the real meaning of Christmas. Organized labor finds itself in a similar position with regard to Labor Day. In a sense, Labor Day is a celebration of a certain bounded group, a celebration that has transcended those bounds and become a national holiday. In doing so, it has changed both in custom and, for many people, in meaning.

How this has occurred is a topic outside the scope of this discussion, but we can compare Labor Day to other large scale identity celebrations that are perceived as more or less political depending on the group involved. Organized (and demanding) labor is threatening to many. Likewise, some may find Puerto Rican Day, Día de la Raza, or Gay Pride Day threatening.

In the United States, St. Patrick's Day has become generalized as a national holiday; I'd argue that it has become a kind of spring carnival (see Santino 1994a), as the Irish have become assimilated. Still, politics having to do with particular issues of Irish and Irish American identity arise, as when, for instance, an IRA leader was asked to be the Grand Marshal of the St. Louis parade in 1997. In that case, Irish Americans were asserting a specific identity vis-à-vis national and international politics, and the result made many public officials very uneasy. Conversely, when organized groups of Gay and Lesbian Irish Americans ask to be permitted to march in St. Patrick's Day parades in Boston and New York, they are denied participation on the basis that their public identification of themselves as homosexual politicizes the parade (Corrette 1996). Apparently some aspects of one's personal identity, such as a taboo sexual orientation, are thought to be polluting enough to cancel out other aspects in the construction of ethnic identity.

The challenge of gay and lesbian groups to established parades that have become normative by staging their own counterparades (as was done in Boston) leads us to another point on the continuum from festival events that are politicized to political events that include festivity: ritualized challenges to dominant, hegemonic events, as described above. Along with gay and lesbian challenges to St. Patrick's Day celebrations, I would include here Native American Day of Mourning ceremonies on Thanksgiving Day, as well as many African American emancipation celebrations that call into question the glib assertions of the Fourth of July by being held in close proximity to Independence Day. Further along the continuum are occasional public events such as political rallies for or against an issue (stop the war, legalize abortion, save the whales); or causes, such as the Million Man March or the Promise Keepers in Washington, D.C., or the Million Woman March in Philadelphia.

More spontaneous social and political instances that involve ritual components and are structured more or less traditionally include spontaneous street celebrations following a sports victory (which, especially in major cities, often become riotous); and the many forms of rough music that include charivari, burning of effigies, tarring and feathering offending individuals, and what Robert Blair St. George terms "ritual house assaults" (Alford 1959; E.P. Thompson 1991; N.Z. Davis 1975; and St. George 1995). This last form of social control was especially prevalent in the British American colonies, particularly prior to the establishment of permanent police forces. Recent studies of the role of crowd justice imply that such traditionalized and ritualized events are prevalent in areas where national identity is contested, as in the British colonies in the late nineteenth century, and in Northern Ireland today. We see a form of rough music in the rattling of litter-bin liners (trash can lids) by Belfast women at patrolling British soldiers, an act that not only insults the soldiers but also warns residents of their approach—the noise is both aggressive and functional.

The rattling of litter-bin liners (trash can lids) on the pavement puts the soldiers in an untenable position—they are being "wrong footed" (Buckley and Kenny 1995). To respond by attacking the women would make the soldiers look even worse to the residents and to the world, but their accepting it empowers the women, at least momentarily. More general uses of provocative symbols follow logically here. The displaying of certain flags at certain times within certain contexts is often felt to be offensive by some, and believed to be done with this purpose in mind. Thus we have Joe McMullen complaining of his neighbor's union flag, while the neighbor would insist it is proper and right to fly the national flag on a national holiday. In the United States we see similar issues surrounding the incorporation of the Confederate flag into the design of the state flag of Mississippi, the use of flag decals on automobiles, or the display of sexist posters at a workplace to which women seek access. The situation in Northern Ireland parallels these cases.

So as we have seen, many festivals devolve into violence, and some are controversial for this very reason. The New Orleans Mardi Gras has been under scrutiny for decades because its carnival threatens to become chaos, although the purported villains who are said to bring about this violence are always outsiders or marginal groups. Over the decades, newspaper accounts have warned of African Americans, out-of-staters, bikers, and hippies bringing trouble to the festival. Even the Fourth of July was a site of both partisan conflict and class conflict in the early decades of this nation. There were many voices suggesting that its observance be discontinued (Travers 1997).

Other, more spontaneous, street celebrations are often rowdy and disruptive, as mentioned earlier with regard to sports victories, the celebration of which have turned into near revolution in some American cities. Likewise, what Coffin and Cohen term the "spring riot" continues to afflict American college and university towns. For instance, the small midwestern town of Bowling Green, Ohio, where I live and work, experiences "Merry Madness" or its equivalent annually. Some years ago, college students celebrated the end of the academic year by congregating on a local thoroughfare called Merry St. As the crowd swelled in numbers it grew in disorderliness. Some people started a bonfire on the street, and fueled it with the pieces of nearby wooden fences they had torn apart. The asphalt itself caught on fire. The fire department arrived to dowse the flames, but students pelted the trucks and firefighters with beer cans. Since that time, university and town officials have done everything they could to prevent a recurrence, but students respond strategically. It has become a game of cat and mouse.

Simply put, the license of liminality is limited and contingent and is itself politicized (for example, police can drink alcohol openly while in uniform at the Boston St. Patrick's Day parade, but gay and lesbian people cannot even participate in the parade). Still, the concomitant factors of masses of people enjoying some liberation from social norms and laws create a potentially riotous situation, even among usually law-abiding citizens. Those students rioting at Merry Madness are otherwise socially and politically conservative. In much the same way that strikes and large-scale political demonstrations are very often festive (for instance, the mass protest in Beijing's Tianeman Square was glowingly described as a festival by the American press until the government began killing the demonstrators), street festivals contain the potential for either directed or undirected violence.

In Northern Ireland, of course, this situation is seriously exacerbated by the presence of warring factions. The violence of the paramilitaries informs the peaceful (nonconfrontational, nonviolent) events, such as when the Twelfth of July parade in Newtownards passes the shell of a bank destroyed by an IRA bomb. This violence also informs society in other ways. For instance, Orangemen reacted violently to the police blockade of their parade route along the Garvarghy Road in Portadown in 1996, and Roman Catholics rioted when the blockade was lifted. Neither of these groups necessarily support the paramilitaries and would not perceive their actions as resembling those of paramilitaries in any way. However, violence as a response, as a bargaining tool—violence as theater—has long been established as part of the repertoire of public discourse in Northern Ireland.

## Resistance from the Right

Having established the question of status hierarchies when examining public display events, and having undertaken a survey of the many types of such events and the many ways in which they can be said to be political and politicized, I must note that the forms of public ritual, festival, and celebration are available to all. Thus, in South Boston, people respond to a parade they feel is problematic by staging their own parade. In the case of Diana's death, however, we witnessed a clash of traditional forms: elite versus popular. Street protest, street theater, and street celebration usually involve popular forms and genres regardless of the political intentions of the participants. That is why, in Northern Ireland, scholars and laypeople frequently comment on the paradigmatic similarities of unionists and nationalists, loyalists and republicans. Belinda Loftus regards the opposing groups as mirrors of each other (although Rolston takes issue with this; see 1991, p. 104). This is because both sides march in parades, carry banners, paint murals, light bonfires, and—at least in the past—have both erected arches. Another way of viewing these actions is to see them as popular genres of public action, each of which, as it is utilized, constitutes a popular style. Regardless of the particular political message, the choice of a popular style to deliver it sets the participants apart from the elite or ruling class—which in the case of left-wing, antigovernment resistance, is relatively straightforward. But in the case of those who take to the streets in order to uphold the progovernment status quo, this popular style frequently leads to isolation and frustration. In cases of such resistance from the right (as I would term unionism and loyalism), this approach allows us to understand the frustrations that repeatedly lead to political impasse and violence.

In "The Planting of the Liberty Tree," E. P. Thompson says of the early nineteenth-century demonstrations against those people who espoused the anti-aristocratic ideas of Thomas Paine:

> Each bonfire of the effigy of Paine served to light up, in an unintended way, the difference between the constitution of the gentry and the rights of the people. "Church and king" actions signify less the blind pogrom of prejudice against an out-group and more a skirmish in a political civil war. [1963:113]

In other words, the protesters were of the same social class and on the same social level as the people they protested against. The events Thompson describes were a kind of resistance from the political right; the groups he refers to were demonstrating in support of the institutions of "church and king," and in support of the status quo that was being questioned by followers of Paine. However, by employing the devices of popular protest such as bonfires and effigies, these groups unintentionally identified themselves as being of a different, and lower, class than the people and institutions they supported.

One form of resistance from the right occurs when individuals or groups publicly protest laws or rulings they find distasteful or immoral, such as legalized abortion. Another sort occurs when, as in the Thompson example, people demonstrate against those who protest the mainstream hegemonic institutions of a society (for example, "church and king"); that is, when people protest protesters. In these cases, the protesters on both sides—left and right—share the same repertoire of public symbols and actions: They all draw upon a shared style. For instance, as student protesters threw blood on the Pentagon in the 1960s, opponents of abortion rights today toss blood on employees of abortion clinics. Effigies of the figure of death have been used in recent decades to protest abortion, nuclear power, war, and the tobacco industry. These and other icons and actions are shared cultural symbols that provide a repertoire of public protest and display.

Resisters from the right frequently find that there is a gulf, a class-based distance, between themselves and those whom they support. The style of their demonstrations dramatizes and sharpens these distinctions, which in turn frequently leads to a feeling of betrayal and abandonment. However, the cognitive dissonance caused by the conflict between loyalty to and betrayal by church, king, politician, police, and the like is accommodated by placing the blame on the individuals who wield the power inherent in those institutions rather than on the institutions themselves, in terms of their structures and ideologies. Rather than viewing the actual institutions these agents represent as fundamentally flawed or unjust, blame is assigned to liberal judges, spineless officials, and duplicitous politicians. In fact, this difference in the perception of moral flaw (the problem lies in the corrupt official rather than the corrupt system) can be seen in the treatment of certain legendary figures of resistance, such as, for instance, Robin Hood and Pancho Villa. It illustrates the difference between personal alienation and political action. As the story exists today, Robin Hood fought in behalf of an absentee king and against those individuals who dishonored their royal offices; while Pancho Villa was a revolutionary who sought to overturn an entire political system.

## Public Protest and Popular Style in Northern Ireland

In Northern Ireland, the style of demonstration (a term that is used for public festive displays as well as political gatherings) is the same among Catholics and Protestants, or, more correctly, among nationalists and unionists. But this does not lead to an identification between the opposing groups, because these performative genres are the media for mutually exclusive messages (Thompson's "civil war").

The anti-gay-and-lesbian movement, with regard to the St. Patrick's Day parades in South Boston, is in some ways similar to the situation of the unionists in Northern Ireland. The people of South Boston are working-class and lower-middle-class (if these terms have any meaning at all), and they have a history of resistance to unpopular laws that directly affect them, such as court-ordered school busing in the 1970s. South Boston is in fact separated geographically from the rest of the city to some extent, and the identification of people with place is very strong there. This is an Irish cultural attribute, and the South Boston population is, ironically from this perspective, overwhelmingly Irish American. During the years of the busing crisis, resistance to the unpopular court ruling took the shape of street theater: People took to the streets simultaneously praying and throwing rocks at school buses carrying African American children. The city government correctly called them lawbreakers for their actions; in turn, residents of South Boston learned not to trust city and state institutions. Indeed, they blamed their problems on the judge who ordered the busing, and who lives in the wealthy suburb of Newton. As with the Northern Ireland unionists, we can see here as well a class-based identification of the South Boston working-class Irish Americans with the working-class African Americans from Roxbury—again unfortunately, an identification that the Irish Americans most vigorously deny and find very threatening. All the same factors of wealth, class, residential location, race, and ethnicity come into play here.

While the content and messages of the South Boston protests were morally reprehensible, the *style* was working-class (and unlawful). The people took to the street and assumed power in a dramatic and ritualistic way, historically consistent with many European and American precedents: as people did at the Boston Tea Party; during the many colonial ritual house assaults; at European charivaris and performances of rough music; and as students did in Harvard Square and throughout the world in the 1960s. Staging these public protest events brought the people of South Boston into conflict with institutions they had always believed in and supported. The effect of this was to radicalize them in a way, but from the right rather than the left. No longer are corrupt or unprincipled individuals to blame; instead it is the government itself. Militia groups, abortion clinic bombers, and the Ku Klux Klan are all examples of class-based right-wing frustration with the refusal of hegemonic institutions to redress people's grievances.

There is little difference here between left- and right-wing groups in that both see themselves as opposed to problematic or intolerable social institutions. However, when the same groups understand themselves to be aligned with these institutions in other circumstances, a sense of betrayal develops. Further, people miss the class-based similarities, manifested in the style of their displays, among themselves and the groups they are opposed to, as against the very real class distinctions between them and those who occupy positions of social and institutionalized power. So, for instance, when gay and lesbian people want to be included in the St. Patrick's Day parade in South Boston, they too use the parade form to make their public statement, and in turn spawn counterparades. In all cases the parade remains the form by which individuals make kinetic and visual claims for authority, validity, recognition, and power (Davis 1986).

The role of the rank-and-file police is interesting here, because they too are generally of the people but enforcers of institutional power. They demonstrate solidarity with the local population when out of uniform, or through large-scale work slowdowns as was done in Northern Ireland by the Royal Ulster Constabulary—an overwhelmingly Protestant police force—in the early 1970s, during a time of massive anti-Catholic violence. In fact, Leigh Corrette has termed the South Boston St. Patrick's Day parade a "carnival for the right," because the quotidian rules are relaxed for those supportive of the status quo. Police in uniform openly drink alcohol during the parade, for instance, but that sense of license and communitas does not extend to gay and lesbian Irish Americans (Corrette 1996).

The protests, demonstrations, and public displays in Northern Ireland and on the streets of South Boston are by and large within the same social class. Northern Ireland is particularly interesting and important because the situation of political struggle (and display) is not restricted to a single organized (if often ad hoc) resistance to a powerful government. Rather, there are two such movements, one opposed to and one supportive of a ruling structure. Both groups are largely working-class and both parallel each other. Northern Ireland may be unique internationally in this regard: It is not simply a matter of a colonial government being met with an organized resistance movement. The situation is further complicated by the existence of organized resistance from the right. As shown in Figure 21, the UFF, UDA, and the entire loyalist and by extension unionist movement identify with the hegemonic power—the queen and the government—but share the style of the movement to which they are opposed. They are similar to the government in sentiment but similar to republicans and nationalists in style. The IRA, republicanism, and by extension nationalism is in opposition to the colonialist government. This is relatively straightforward and undisputed. Both the government and the oppositional groups understand this. But the unionists and loyalists are conflicted: They share sentiment with the overarching power structure but share style with their working-class enemies. Members of that powerful hierarchy would never use the traditions and styles of popular protest, and are in fact embarrassed by them. While they may appreciate the political (electoral) support of unionists, they often

### Figure 21

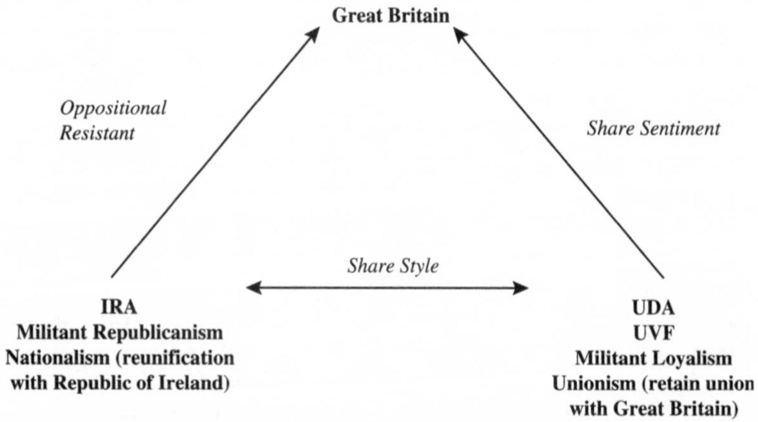

| Great Britain |
|---|

Oppositional Resistant

Share Sentiment

Share Style

**IRA**
**Militant Republicanism**
**Nationalism (reunification**
**with Republic of Ireland)**

**UDA**
**UVF**
**Militant Loyalism**
**Unionism (retain union**
**with Great Britain)**

**Figure 22**

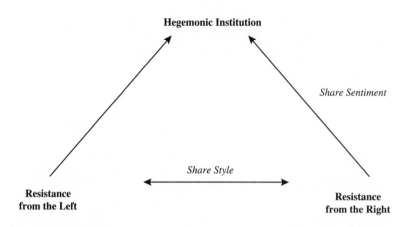

find their class-based, stylized, expressive behaviors distasteful. It might be possible to generalize this principle to right-wing resistance groups in the United States and elsewhere (Figure 22). Such groups experience conflict and cognitive dissonance in the fact that they are despised by the very powers they support—despised based on their social class as expressed in their popular style of public display and protest.

I am leading to a formulation of culture as style of behavior. This formulation sees cultures as both larger and smaller, including what are usually referred to as subcultures. In this way the concept of style as socially constructed behavior can be used as an entrée into the study of various cultures or groups of people. In the examples cited above, the opposing groups have in each case stylistically and culturally more in common with those with whom they clash in the streets than with those who maintain the institutions they are supporting.

# CHAPTER SIX

# Shared Style and Paradox

Twice a year, the city of Derry (or Londonderry) is the scene of major celebrations. In August, the Apprentice Boys—a fraternal organization named for the apprentices of Derry who held the city gates shut to prevent Governor Robert Lundy from surrendering to the forces of King James—commemorate the lifting of the siege. In December, Lundy is burned in effigy. The summer demonstration is particularly controversial because the Apprentice Boys have a reputation for rowdiness in their parades. They are said to make a mess, littering the streets with beer cans and food wrappers. Worse, they are associated with public urination. The leaders fiercely deny these accusations, explaining that with such large crowds of participants there is bound to be some of these activities, and that they cannot control everyone involved.

Unfortunately, I would have to say that my own impression of these occasions matches the unpleasant descriptions. I have witnessed a good deal of public urination and occasional vomiting in Derry, much more so than at other events of this type. When I mentioned this to a Roman Catholic friend with whom I was staying, he pointed out that Derry is a majority Catholic city. It is frequently mentioned as an example of successful power sharing in the city council. The urination, he said, was a political statement: "It's a Catholic city—piss on it."

Whether his perception matches the festival participants' intentions or not, the public urination, while it can be seen as inversive, can also be seen as a demonstration of ownership, disdain, and power.

Interestingly, these events exemplify the dimensions of commemoration and performativity quite well. Officially they are commemorations of seventeenth century events. But each year the Apprentice Boys face controversial decisions and demands that are very much a part of the here-and-now. For instance, like so many other parade routes in Northern Ireland, theirs has been challenged—including their right to walk on the medieval city wall. The Apprentice Boys insist upon this right as part of their heritage, with various results each year. Sometimes they are legally denied access to the wall, sometimes they are allowed to walk a small section of it early in the morning, and so forth. The population of Derry is sharply divided by neighborhoods, and the Catholic population complains that the parade takes place in their full view. The Apprentice Boys claim the right as citizens of the United Kingdom to walk the Queen's—and their own—streets. So while the parades commemorate highly significant events of the past, events that led to the slogan "Never Surrender," they are also performances that address contemporary political and social concerns, with a view toward affecting them.

## Carnival and the Carnivalesque

As I was watching the parades and walking the streets of Derry, thinking that in fact many of the participants did seem to be rather trashing the city, I was reminded of Mikhail Bakhtin's concepts of carnival and the carnivalesque. Bakhtin explained the grotesqueries found in the work of Emile Rabelais as carnivalesque humor, derived from the popular culture, or folk culture, of the medieval carnival itself, and similar celebrations. Despite a rather strict definition of carnival, Bakhtin includes events as disparate as cattle slaughters, grape harvests, papal jubilees, and the Feast of Corpus Christi within the rubric of the carnivalesque. The humor of these festivals, as he describes them, was inversive, exaggerated, and impolite; but more specifically, this humor referred unashamedly to the bodily functions of "the lower stratum" (1984[1965]:20), both sexual and excretory. In opposition to polite, rational, cerebral rule-oriented (and thus power-oriented) institutions of the quotidian government, the popular carnival was full of references in costumes and in verse to the phallus, the bladder, and the anus. Bellies were grotesquely extended as with food or with child; noses became phalluses. For Bakhtin, the term "carnivalesque" means the specific use of this kind of symbol, this kind of language, these kinds of dynamics. The term enjoys wide use in scholarship today, but its usage has been modified to the point where the Bakhtinian references to the bodily stratum have all but been forgotten. "Carnivalesque" is now frequently used simply to mean festive.

So I wondered as I wandered—this urine, or this splotch of drying vomit on the sidewalk—are these carnivalesque? Certainly they are inversive of the rules and decorum of everyday life, and obviously they are of the (lower) bodily stratum. However, Bakhtin describes carnival, ideally, as a "pure form" (1984[1965]:30). Has such a pure form ever existed in reality? Bakhtin has in mind the parody linked with the impolite. The Bakhtinian ideal (communitas and an essentially comic view of life and the world), this does not describe the Apprentice Boys' parades. These are too much a part of the political conflicts of their time and place. Bakhtin, in fact, characterizes parades as state-sponsored attempts to control the carnivalesque spirit (1984:33). The Derry parades and Orange parades reinforce hierarchy while suspending some prohibitions. While they partake partially in the carnivalesque, their intention is to inscribe hegemony upon territory.

Moreover, the Orange Order and the Apprentice Boys officially look down on "carnivalesque" elements that occur during their parades; they close their eyes to them, and officially condemn them. To this extent, the bodily waste of the rank and file, the younger members, and the members of the "kick the pope" bands often affiliated with paramilitary groups, *is* a carnivalesque response to the stuffy control of the established powers-that-be in these organizations. However, Catholic residents read these actions as bodily political assaults on their home.

## Festivity, Inversion, and Power

Festivity, inversion, and the carnivalesque are not synonymous. Not only is every festive event not necessarily carnivalesque; neither is it necessarily inversive. Even if it is inversive, it may not necessarily be resistant or oppositional. The Northern Ireland public ritual genres are a subset or manifestation of carnival in the broader sense, not in the more specific reference to the European pre-Lenten Fastnacht carnivals or African-Caribbean events such as Mardi Gras in New Orleans or Carnaval in Rio de Janiero. They are more carnivalesque than, say, public religious processions, but much less so than carnival itself.

However, even given the tumultuous, world-upside-down frivolity of the New Orleans Mardi Gras, conventional wisdom regarding the Bakhtinian functions of carnival can be challenged there as well: Many of the participants—the Mystic Krewe of Comus, for instance—are drawn from the city's elite social aristocracy; from the ruling class, as it were. Joseph Roach has shown how the founding members of these krewes had been former Confederate soldiers, which is not surprising, but also members of the Ku Klux Klan (Roach 1996:261–262). Today, much of the Mardi Gras public ritual display goes toward reifying the status quo, as does, certainly, the concomitant private indoor displays of wealth and power, the debutante balls and masquerades. The idea that carnival is always subversive is clearly challenged by such phenomena. The New Orleans Mardi Gras consists of events from the bottom up (such as the krewe of Zulu, the Mardi Gras Indians, and the street crowds) as well as from the top down. Sometimes these clash dramatically and theatrically, as when the Indians' parade route crosses that of Comus. Still, both the elite groups and the African American groups, along with middle-class and suburban groups, contribute by their actions to the overall festivity and communitas that characterize Mardi Gras and, indeed, New Orleans itself during the carnival season. (See Gaudet 1998 for an example of Mardi Gras as normative behavior.)

The fact that the clash of class privilege is manifested in parades is important. The kinesthetic, invasive, mobile, and linear qualities of parades are more about generating meaning than defending or maintaining it. The places through which parades move may be replete with their own semiotic signifiers, but parades are themselves creative and generative (de Certeau 1984). Leigh Corrette's concept, "carnival for the right" (Corrette 1996), is especially relevant to the Orange parades in Ulster. Peter Burke points out that in Medieval European carnivals, despite cross-dressing by the men, women have historically remained on the sidelines. The world has not been turned upside down; rather, these were rituals of masculinity (Burke 1978). Indeed, as in the present-day Spanish Tomatina, where festival participants smash each other with tomatoes (and tear off each other's shirts) we see demonstrations of heightened masculinity rather than true social leveling at many "carnivalesque" events.

The public rituals of presentation by the Orange Order and the Apprentice Boys of Derry are also carnivals of the right, and like the elder Mardi Gras krewes, they ritually uphold the status quo. The rowdyism that exists, for instance in Derry, where the public drinking results in public urination, can be seen as simply another demonstration of power.

Unlike the Mystic Krewe of Comus, however, the bulk of participants in the many parades, the bands, are not made up of the social elite, although the leaders of the Orange Order have controlled—and to some extent still control—much of the political power in Northern Ireland. Nevertheless, most of the participants are very much outside the domains of power, as middle- or working-class citizens, although as Protestants and unionists they may enjoy certain privileges Roman Catholics still do not (although many Protestants would strongly object to the truth of this statement). Indeed, the Orange Order symbolically represents itself as a "chosen people" (Buckley and Kenney 1995:173–193).

So festive events in New Orleans might be called carnivalesque, but they do not necessarily perform in the ways Bakhtin describes carnivalesque events performing. And such events may not necessarily invert, even temporarily, the status quo. Don Handelman has emphasized the point that inversive events are always set aright after the carnival is over, and that this is part of the statement (Handelman 1990). They are not intended to effect permanent change in the way that rites of passage do, or that strikes do. Indeed, the social hierarchy of power and privilege may in fact be reified by the performance of festive and carnivalesque events. This is the case in

Northern Ireland; it can lead to frustration, as I suggested above, concerning the identification of such events with the power elite while they reflect a style that is a signifier of a different class. The clashes of Orange Order paraders with soldiers and the RUC at Drumcree in the late 1990s exemplify the uneasy relationship of the paraders with the imperial power in London. These parades are supposed to be rituals of presentation (Handelman 1990) in which the display of power relations is naturalized, presented as self-evident. However, this hegemonic bid is challenged in Northern Ireland by almost half the population, and is undermined by high-level political decisions over which the parade participants have no control. When they are denied the option of parading, the very power of their parade is visibly undermined. As a result, members of the Orange Order riot, battling the police and the soldiers who are representatives of organizations and institutions they support. And in so doing, the Orange Order members present a picture of themselves, to England and the rest of the world, as indistinguishable from their adversaries.

## Concepts of Tradition

Events such as the parades and confrontations on the Garvaghy Road in Drumcree become part of the repertoire of shared memory. They provide new or modified contexts for future similar events. An unfortunate example is the 1998 death by arson of three young children during the Orange Order's standoff at Drumcree, a sobering event that took the wind out of the sails of the confrontation. Certainly it will be remembered and used as a way of viewing future Orange parade demands. The same is true for the violent events of the two previous years. Likewise, the killings at Sean Graham's Betting Office, and their subsequent memorialization, have taken on a life of their own—as the parade route on the Lower Ormeau Road is contested by residents, as a permanent commemorative plaque is installed, as other events coalesce around it. Loftus says that every time a mural is painted or seen, it recapitulates past usages and simultaneously creates a new context for future memory (Loftus 1988). This is true of all the artifacts and events we have examined. We experience them in terms of others we have seen, yet the new experience of a particular event becomes a memory, a context, for the future.

At various times throughout this book I have referred to the parades, murals, and shrines as traditions. The concept of tradition is invoked quite often in Northern Ireland as a justification for parade routes, for instance, or to explain the use of a Lambeg drum—"It's not to intimidate, it's just our tradition." Indeed, the British and the Irish cultures in Northern Ireland are referred to as "the two traditions."

Dan Ben-Amos (1984) has identified at least seven ways in which scholars use the term "tradition," and there are probably others. Henry Glassie (1995) has defined the term as the use of the past in the present to effect the future. I would argue that, when speaking of tradition, people sometimes refer to form, sometimes content, and at other times paradigm. For instance, turkey is traditional at Thanksgiving in the United States. So is the meal-as-feast at which it is served. Many vegetarians keep the form—the meal—while altering the content by substituting a non-meat main dish. When we discuss form, though, it leads us to a consideration of paradigm: in this case, that feasting is one way of marking special occasions. In Northern Ireland, it might be said that on one level the traditions are in the details: the colors green or orange, the Lambeg drum, the orange lilies, the shamrock. On a more abstract level of analysis, the traditions are in the form or organization of those details—so that all interested parties in Northern Ireland can be said to share the "traditions" of wall painting and parades. Finally, the groups tend to share paradigmatic structures or cognitive models for social action such as fraternal organizations, as manifested in their social organization and formalized responses to social problems.

Tradition, so named by its participants, is self-conscious. The use of the past in the present is done knowingly and is often framed as such. The "pastness" is marked, placed in the foreground by such actions as carving the turkey with the family silverware, restoring favorite decorations annually, or recounting the narratives of previous holiday occasions. If the past were not invoked, the term would be entirely identical to the term "culture." We eat turkey on Thanksgiving because we have always done so, and because others we know do. Those who do not are aware that they are in some way unusual, and often call attention to the fact of their nonconformity through some act of parody or inversion. As I have said, however, people can act traditionally on the levels of content, form, and paradigm.

For instance, I have written elsewhere (1986) on people's public decorations of their homes at holiday times. The various decorative elements that are juxtaposed in these public decorations are mixed in origin: some are storebought, some homemade; some are organic, while some are made of synthetic materials such as plastic. Rather than isolate the hand-carved pumpkin from the plastic jack-o'-lantern, terming one folk and the other mass culture, I considered them all together as a single artistic, decorative assemblage. It is the act of decorating, along with the symbols used, that is here traditional. These decorative holiday assemblages are a recent development (although with many antecedents). Thus paradigm—marking seasonal time with decorative display—and aspects of content—images of harvest and death—are here traditional, while the form (assemblage) is emergent.

This is somewhat related to Hobsbawm and Ranger's ideas concerning "invented traditions," although this term too has suffered from overuse and confusion. Hobsbawm and Ranger seem to mean, on the one hand, that all traditions are invented rather than superorganic; and on the other hand they refer to traditions that are wholly invented at some particular, documentable moment of time, or else are highly interpreted and manipulated by a ruling class, thus entering the idea of power relations into the discussion (Hobsbawm and Ranger 1983).

Take, for instance, the case of rough music (see Alford 1959, Davis 1975, and Thompson 1991). In Northern Ireland we have seen numerous examples of this—for instance, litter-bin lid rattling to annoy British patrols, to object to their presence, and to warn of their approach; or the making of a racket as a parade passes a Roman Catholic church. Just as scholars have a categorical name for these actions, so do we see them as being within, or of, a tradition of similar or related actions that have preceded them. Scholars see them as paradigmatically traditional. The participants would recognize the form and the content—we beat the drums or rattle the litter-bin lids under certain circumstances—but not the paradigm as such. They would probably not recognize a connection between these actions and a wedding charivari, for instance, or even between the pounding of the drum and the rattling of the lid. Certainly they would not connect these to effigy burnings, tar-and-feathering, or ritual house assaults, as scholars routinely do.

Robert Blair St. George has written on ritual house assaults in colonial New England (St. George 1995). He demonstrates the symbolic logic of the activities, the patterns that underlie and determine who and what is attacked, and how it is attacked. Even the route the reveling demonstrators take is shown to be symbolically loaded, as they intentionally pass certain public buildings and the homes of key officials. In recent years I have encountered several interesting cases of ritual, or symbolic, house assaults. While some of these may not have the same universal currency as did the colonial cases, an example of one will serve to put the idea of tradition in perspective.

A professor from my university woke up one morning to find that the façade of his house had been covered with sanitary napkins that had messages written on them in red ink suggestive of menstrual blood. Some of these messages were quite hostile; one read, "Suck my bloody cunt." The professor is well known locally as a very outspoken, notoriously right-wing individual who refers to the women's movement as "feminazis." He is very controversial in class and around campus, and he loudly opposes feminism, affirmative action, gay and lesbian rights, and so forth. The perpetrators were a group of women who clandestinely put up messages around campus to challenge people's complacency, as they see it. On one such night they got the idea to "pad" this professor's house. They purchased a quantity of adhesive sanitary napkins, got together, drank wine, and wrote the messages. Then, under cover of darkness, they drove to within two blocks of the man's house, walked the rest of the way, and stuck the pads to the front of the building. While this was done secretively, the results were highly visible for the inhabitants of the house, as well as passersby, the next morning.

I call this a ritual house assault, related paradigmatically to those described by St. George. The participants themselves, however, do not see either the content (the "bloody," theatening pads) or the form (the attack on the house as symbolic of the person who lives in it, leaving an embarrassing public sign of one's censure) or the paradigm of ritual house assault (or what Dorothy Noyes [1995] has termed "façade performance"). They certainly understood and capitalized on the powerful symbolism of menstrual blood, and they assumed that the target would understand that as well, in view of his position on women's issues. But they did not view the act as traditional—then. Since then, they talk about "padding" others. The initial padding has itself become a paradigm for action, and they do recognize this as traditional.

To say that tradition involves conscious choice and recognition is to say that tradition on the level of content and form is an emic, in-group category, and that it is intertextual and reflexive. This is how the parades and murals and shrines work in Northern Ireland. To talk about paradigmatic tradition is etic and analytical (Ben-Amos 1976) and leads one into thorny difficulties in distinguishing tradition from any other aspect of culture. These three levels of form, content, and paradigm are each subject to analysis. All signify. Each has its own concomitant level of signification. Belinda Loftus has made some very useful suggestions regarding visual symbolic forms in Northern Ireland that can be applied to each of these levels, particularly content. Paraphrased and condensed, she says that any specific visual image is the product of a maker or group conditioned, though not completely determined, by social, political, economic, and religious factors, personal context, technology, location within institutional structures, and its use of existing visual conventions; that the private and public significances of visual images are intertwined; that an image or meaning is not static, but is further developed each time it is used or reproduced, and that in the course of these processes it both acquires additional layers of meaning and has a real impact on social, political, economic, and religious developments (Loftus 1988). We have seen these dynamics at work throughout this book, particularly the idea that each new use of an image or a symbolic artifact recapitulates past usages while also adding a new nuance to the image; and that private and public significances are intertwined, as with spontaneous shrines. We need to regard the multiple meanings and usages, along with the multiple populations, identities, and class levels, as co-occurent.

## Border or Box

If the various images and genres such as murals, parades, shrines, and so forth, cannot be divorced from each other in daily society but are experienced in conjunction or disjunction with each other, then so are the various people, with their jumbled, complicated identities, who are involved. For instance, there is in Belfast a 20-foot-high steel wall that divides the Catholic Falls Road area from the Protestant Shankill. It is called the Peace Wall. It is extended through much of Belfast by highways that are difficult to cross. In its entirety, it is known as the Peace Line. Residents on either side of the wall fly their flags high enough to be viewed by residents on the other side in another "up your nose" move. At points where vehicular traffic has access to cross, children sometimes throw stones at children on the other side. This is a border. It divides. It creates two populations, them and us. It is tempting and easy to view the overall situation in Northern Ireland in dualistic terms, employing the metaphor and rhetoric of the border. Sometimes it is useful, as in this case when we have so clear a bifurcation. Still, it is worth noting that I was informed by a republican resident that despite the official name for the wall ("The last wall left standing in all of Europe, now that the Berlin Wall is fallen," he said), he considered the real purpose of the wall to be containment and surveillance. There are several levels of power and complexity in Northern Ireland, and a "two tribes" view allows observers (such as the MPs described above) to disinvolve themselves, distance themselves, and abdicate responsibility. The "crazy people killing each other" stereotype not only is useless to understanding the problems, it is also disingenuous. While there is inescapably a dualistic and dialogic aspect to these people and the public events we have been examining, the forms and the people can be seen as engaging each other, not so much across a political divide (as with the flags at the wall) but within it.

Ulster is a place with dual, and dueling, histories, two national memories. Two mutually exclusive master narratives compete, but there are many shades of variation within those two camps, different religious and ethnic backgrounds, with allegiances that shift over the years. The border metaphor is of limited value, and is not accurate. Let me illustrate with the following example. After a class I teach, I am in a bar with a group of students. Some are women, some men. Some are white, one is Asian, one is African American. Some are gay, some are straight. Some drink alcohol, some do not. I am the only professor; the rest are graduate students. "Boundaries" exist, or could be constructed, along each of these dichotomies, but at any given moment all those factors coexist. Further, they will each be more or less relevant at any given moment. Also, the dichotomies are simplistic: since there is an Asian and an African American present, we could describe a white/nonwhite dichotomy, but both the African American and myself are from the United States and are both male; the woman is Korean. We could describe drinkers and nondrinkers, but those who drink beer make jokes about those who order mixed drinks, not about those who order soft drinks. Viewing the situation in terms of dyads is not productive. It is only in particular situations, when any of those identities are foregrounded, that they are relevant, such as when a man refers to a mixed drink as a "girl's drink" (that a man has ordered), or when they ask me about a forthcoming assignment, or—more formally—at events such as a gay pride day celebration.

Likewise, in Northern Ireland, even within the unionist "community," (a term that in fact refers to a large and differentiated group), there is diversity due to (among other things) socioeconomic standing (class), political allegiance, religious and ethnic background, age, gender, residence, and so on. These factors influence political sentiment: Men more than women are actively engaged in paramilitary activity (but see Aretxaga 1997), as are working-class more than middle-class, younger (and perhaps unmarried) rather than middle-aged. The forms we have surveyed mirror these distinctions. The murals often speak to paramilitary concerns, as do the "kick the pope" bands, both male-dominated, but these latter are more or less successfully held in check by older (and more "respectable," that is, middle-class) members of the organizations. The shrines are frequently but not exclusively initiated and maintained by females.

Do we have a dialogue here? Is there competition among the groups for position, for the ability to spin the message? Yes, to some extent, among all the various factions. There is more to this than two neighborhoods on either side of a wall. We are not talking about borders here. To remain with the spatial metaphors, I see Ulster as a box, analogous to the depiction of the classic paradox:

**Figure 23**

All statements within this box are true
I love you
I hate you

Similarly, Ulster is a box wherein the struggle to name the land and control the master narrative is material:

**Figure 24**

All these statements about Ulster are true:
Ulster is Irish
Ulster is British

Buckley and Kenney (1995:x) say that the people of Ulster have been dealt a difficult hand. I say they have been born into an impossible situation, a paradoxical box in which the values and beliefs of each group are in direct contradiction to those of the other group. Ulster—more correctly, Northern Ireland—is the box. There are borders, of course—between north and south, most dramatically, with the border patrols and the border towns that are hotbeds of paramilitary activities, and at which illicit as well as legal crossings occur at security checkpoints; and many of the expressions of the division are linear and dualistic, such as the Peace Wall. However, to me the most useful metaphor with which to apprehend the complexities of Northern Ireland is a spatial one, areal rather than linear, with multiple competing groups. There is dualism: Ulster is a space within which two self-defined peoples, two cultures, have to coexist. And despite all of their (often unacknowledged) similarities (they both drink the same Guinness), they consciously maintain differential expressions (sports teams, names, flags, political allegiance), marry endogamously, and so on. Still, other peoples with different histories and backgrounds live together without the kind of strife seen in Northern Ireland. The crucial factor here is that the worldviews, the social beliefs, the most basic, fundamental views of the world in which the two groups live are contradictory. We are not talking simply of difference here, but of mutual exclusivity. Ulster is either British or it's not; it is either Irish or it's not—they cannot both be right. In this sense, they in fact cannot coexist. These are differences as fundamental as the dirt itself.

Henry Glassie has explored the deep semiotic meanings of dirt in his study of Ballymenone in County Fermanagh in Northern Ireland (1982), and Seamus Heaney has discussed the way the Irish landscape is infused with myth, legend, and history. Ulster—the tangible, physical Ulster, the Ulster that you can feel with your fingers—is Irish, Celtic, Catholic, nationalist, part of Eire. Ulster—the tangible, physical Ulster—is British, Teutonic, Protestant, unionist, part of the United Kingdom. The two sets of concepts cancel each other out, just as the paramilitaries attempt to cancel each other out, to erase each other's history and thus each other's culture. The people of Northern Ireland are accused frequently of being lost in history, but the truth is that they use history to talk about the present. This is done both literally and figuratively. People feel a real sense of historic grievance, and history is used metaphorically to talk of the present, as they do, for instance, with the concept of the siege. These events are captured and transformed in murals and banners, and, as Loftus says, each use creates a new context to add to the ever-changing interpretation of past events in the present.

According to one perspective, in the box that is Northern Ireland all things are, or should be, Irish; according to another, all things are, or should be, British. Both sides claim they are perfectly willing to tolerate, to a greater or lesser extent, the "tradition" of the other, but with the condition that the other must recognize its own lawful, rightful ascendancy. Both sides sincerely believe their positions to be morally, ethically, and legally correct. The debate is seen as being about true essences as opposed to social constructions. The existential nature of the very land the people stand upon as they debate is in contest. One position cannot be validated without denying the other; it is a zero-sum game writ large.

Of course, we see it as entirely socially constructed, but that is an easy position to take from the outside. The genres, forms, and activities examined throughout this book are the ongoing means and process of social construction. Often they enact, dramatize, and create the paradoxical division. Those who died at Sean Graham's did not choose to do so; they chose neither the time nor the place of their deaths. But the shrine that Eileen McManus chose to initiate there was seen by all those who used this major street, one that is also part of a parade route for the Orange Order. They have walked past this site for years but now they are met with deliberate resistance, due to the events at the betting office and the population of the neighborhood. The general issue is that Orangemen can parade anywhere, while nationalists are restricted. Here it is specified that Orangemen can use the Sean Graham's-Ormeau Road route. One group accuses the other of insensitivity, and demands the right to control such events in their own neighborhoods; and the other group responds that they can walk on any public street in their country, that they refuse to recognize "no-go zones."

## A Discourse of "Race"

Northern Ireland is intercultural, interreligious, interChristian, really. But it is also, of course, interpolitical. To the extent that the people define themselves and perceive themselves as British or Irish, it is international; and to the extent that they define themselves as Teutonic or Celtic, interracial. In fact a racial discourse has been constructed around the troubles. Promoting contemporary socially constructed and socially generated identities is a genuine, if imagined, racially based discourse. Contemporary social distinctions are naturalized, constructed as genetic, as biological, as inherited. As one man once said to me, "It's in your balls."

Viewed from this perspective, the development of the Cruthin history is seen as the creation of a racial history that explains and legitimizes a contemporary people (see Adamson 1974). It reverses the rhetoric of the "native Irish." In this particular invented history (remembering that all histories are invented), the Cruthin are more native than the Irish. They were there first; the Irish are invaders from the south, and the Scottish planters were merely coming home after centuries of exile. Also, this history posits a third race, neither Celtic nor Teutonic, that is the wellspring of Ulster culture at a time when Great Britain seems to have abandoned Northern Ireland to the rule of the Irish, a trend the Ulster British abhor. Ulster as an independent state, like Articles 2 and 3 of the Irish constitution (which claim dominion over the six counties), is more powerful as political rhetoric than pragmatic policy, but it is appealing to many people in its symbolism. The concept of race that this history is based upon is scientifically questionable. The Cruthin theory is highly suspect. But it is a social rhetoric that erases the social construction of political reality by naturalizing it.

## Territory

On the elaborately painted walls on the Lower Newtownards Road are the words "The Ulster Conflict is about Nationality." Rhonda Paisley says, explicitly, that it is about territoriality. In so doing, she discusses many of the visualizations we have been considering throughout this book, and she articulates their relationships to this territoriality and to the fact of war itself. She begins by talking about painted curbstones. Interestingly, in a conversation with me at a later time, Ian Paisley would cite curbstone painting as an act of aggression. In Ms. Paisley's words:

> The paintings of curbstones and the visual side of what's happening, I think that did die down considerably, and it has come back into vogue during the years of the Troubles, the actual paintings of murals and such. Originally it was really the Orange that painted the murals coming up to the Twelfth, and they seemed to be the first who did that, and then that sort of faded. And those murals tended to be touched up every Twelfth, you know, and it was of King William. But then the whole hunger strike thing started up the mural painting again. And I think the use of the Tricolour became very much more a territorial thing in the province, a statement of "This is our state" or "This is our area." And I think the union flag became used similarly again, it wasn't just put out as a national flag at a time of festival, it became, you know, a territorial thing, and I think it has continued, and even more so.
>
> I remember even from my childhood days, as you drive around Northern Ireland you would see more Tricolours flying now than you would have ever seen years ago. So I think that there definitely is a territorial marking, like, a visual marking of territory. I think the situation in Northern Ireland *is* a war over territory. I think that Articles 2 and 3 of the southern constitution very clearly say that, you know, their claim is to jurisdiction over the North, and I think that people in the North who don't view a united Ireland as what they wish for the future feel that it's a claim to their territory, and it is a war about territory.

I asked if it boiled down to the Republic of Ireland claiming the six counties and Great Britain saying, "No, they are ours." She replied:

> I think it's more complicated. I don't think Great Britain is saying, "No, they're ours." And I think that's why the people here are indicating to Britain that we wanted them to stay here. And I think there's a sense that Britain isn't being forthright about their territory here, because some people feel they are letting them go slowly, rather than just coming out and saying "We're letting you go." And that's the great fear, which is creating the situation and making it even more intolerable. For example, last week there was a situation where the fishing rights were being disputed between France and Jersey, and on the news there was a big report. The British were there to make sure the union flag was not taken down and the French flag was not put up in Jersey as a claim to the fishing rights. And then you look at what's happening in Northern Ireland; the whole Flags and Emblems Act was changed to make sure that flags and emblems could be flown that belonged to another national country. Another national emblem of another country. And you know, I think. there is a contradiction there in what Britain is prepared to do in other places.
>
> To allow everybody to fly the flag they are most comfortable with, that's not dealing with the problem. I don't think anybody can object to someone flying a flag or using a symbol or an emblem that they feel personal loyalty to. A lot of it is the way in which it is done. No country can have two national flags. Not whenever there is a declaration that one of those governments wants to take over the territory of the other one. That's where it is a declaration of—ah—war, if you like. And that's where it becomes a problem. I mean I can tolerate someone feeling certain loyalties to a flag or an emblem that I wouldn't have loyalties to. I think that's your civil right. And I would defend your right to do that in Northern Ireland, and that's okay when there's not another claim to another country's jurisdiction. That's where it becomes a problem.

Rhonda Paisley provides an insider's perspective, an emic explanation of what can appear trivial to those on the outside, and she explicitly connects public display to acts of war. On several occasions I found myself asking people, like so many others have asked, "Why not simply let people fly whichever flag they choose and be done with it?" Because, she points out, such an act is treasonous, certainly to her and to the overwhelming majority of the Northern Ireland British. Moreover, she pinpoints the feelings of isolation and betrayal these people feel regarding the British government's policies. The feeling of isolation from the institutions one is supporting leads, as I have demonstrated above, to feelings of frustration, possibly outrage, and perhaps to violence. Underlying the example Ms. Paisley recounted above is the implication that ultimately these policies are based on economics, rather than on such principles as loyalty or morality. Great Britain is not about to give up fishing rights; that is entirely an economic consideration. However, Great Britain has renounced both economic and strategic interests in Northern Ireland in the Anglo-Irish Agreement, an act that angered and frightened many. With no economic interest in Northern Ireland, events such as the fishing dispute show that Great Britain will have no motivation to retain the union with Northern Ireland, and the republican paramilitaries will provide plenty of reason to dissolve it. Thus the unionists perceive the strategy of violence as having led to victory, and consider their sense of honor and morality severely violated.

From the very tempered and insightful words of Rhonda Paisley we move to a similar level of reflection by Mari Fitzduff, of the Belfast Community Relations Council, whose job it is to bridge the prodigious gaps we have documented throughout this book. She points out that one potential strategy for doing this would be to try to create neutral but mutually acceptable places for large-scale celebrations. Another approach is to attempt, in her words, to "celebrate the lot." The CRC has chosen the latter strategy, in contrast to the Fair Employment Commission, which instead tries to ensure neutral working environments. Ms. Fitzduff says, "In places like a university, for instance, we would rather see celebrated all cultures that emanate from here, and so there's an enormous amount of parading and the question is how you deal with it. Do you try to sort of keep the streets clear or do you try to ensure tolerance, a greater tolerance towards the different groups that want to parade? We certainly go for the latter."

She sees great strides made by the Orange Order in recent years in its efforts to avoid causing offense among Catholic residents, even going so far as to phone local churches before parades to determine the times of their services and thus avoid disturbing them. Still, there are problems, particularly in the Orange Order's refusal to reroute its parades. In Mari Fitzduff's words:

> The fact is that a change in demography means that, for the most part, Protestants have been ousted out of their areas in greater numbers than Catholics, and therefore, historically they are ending up with routes that used to be Protestant but which are now Catholic, and defiantly saying, "We won't accept this change in demography and we will continue to march where we have always marched." There's an enormous amount of variations within the Order, in terms of perceptions of the Order. Because there are a lot more Orange parades they're seen as more threatening, and because it is much more normal for Orange parades to march in areas that are nationalist than for nationalist parades to march in areas that are unionist ... You will often get a contradiction between the top level and the bottom level. Essentially power rests with the local lodges, and even they find themselves curtailed because

they may have a group that meets once a year, but they are joined then, once a year, by these bands that can be quite aggressive and can be very explicitly sectarian, and very often offensive.

But I do think that in the Orange Order some of them have to do quite a bit of reflection of where they are, and they vary. I mean there are some that are more than happy to have their premises used for mixed playgroups and mixed history groups, but they don't go very public about that because there are quite a few people within their own lodges who actually create trouble about that, and would be very threatened. But it's interesting if you watch the news. They have very negative coverage now of the Orange Order this year, but in a sense it wasn't any explicit thing. It was about, "We're forced to stay in our houses on such and such a day." [Indeed, screens are often erected at housing estates, literally locking people in—and out—during parades.] I can see the same resentment in terms of other celebrations. We had a Protestant group complaining that they were going to be locked out of their homes because of celebrations that were going on in the Catholic cathedral there. But they are likely to be lessened because people are likely to be less resentful of celebrations of their own community than celebrations of another community.

In other words, it is more problematic when a parade route traverses the territory of the other population. In regard to Mari Fitzduff's point about the nature of the media coverage, most unionists are very wary of reporters. They do not believe they have been treated fairly or evenhandedly. And as regards the coverage of the July Twelfth parades, a man once told me that in the past the parades were televised continuously through the whole day. From that extreme, he said, it has gotten to the point where they barely even mention the locations of the six official parades in advance of the Twelfth. This man considers that lack of acknowledgement both a slight and a challenge. The more the media downplay the Twelfth by trying to ignore it, the more he is motivated to get out and support the parade.

Turning to the issue of territoriality that Rhonda Paisley focused on, Mari Fitzduff has some interesting perspectives. She says,

In a sense life has been extremely difficult for the unionist communities—in fact, since the civil war started in 1969. In a sense there have been different focuses for their angers and their energies, and a particular focus over the last few years—before that it was the Anglo-Irish Agreement—the last few years it has been Articles 2 and 3 of the constitution. If you look at it, in fact, there has never been any suggestion that there is actually any explicit threat or any intended threat or anything else. This is a leftover of the Irish constitution that they certainly don't intend to invoke, but it is used as a way to—on which to hang their fears. Flags would be a part of it. At one level they are saying the cause of all our problems is the threat from the Republic to take us over, which there probably is no intention of doing. But at another level, I know them well enough to know that if the threat was gone it will be another threat. They feel threatened, and in a sense they will find another focus.

I mean there were people who would see an enormous threat in a Tricolour, and in fact a lot of the security forces would have seen flying a Tricolour as being subversive until fairly recently, but they've learned a lot. It's been quite some time since police have gone into nationalist areas and taken down Tricolours. They have learned to live with it, and it's quite interesting. I was talking to some army people there recently at their training camp, and they actually have a flag which includes both the red hand of Ulster and the Tricolour. To make sure the army understands there is a respect for that aspiration even if it's not a legal reality. There are some people who would say in fact that the aspiration is what many Catholics want, and not the reality of a united Ireland.

In fact I have been told precisely this by a close friend, a nationalist. He said that he was completely against the dropping of Articles 2 and 3 from the Irish constitution, because he thought the threat, the claim, was productive in moving things forward for Catholics socially. However, not only would he not want to see the tanks roll in from the south, he also is quite content with the union as it is. He enjoys the British health care system and does not relish the idea of living in a society in which the Catholic Church's morality is a constitutionally imposed legality. Mari Fitzduff articulates this position, its whys and wherefores, succinctly:

It's partly the idea that it has been the threat of a united Ireland that has actually achieved some of the redress in terms of balances of justice. So that in a sense you wave this particular bait in order to achieve this, but what most— and the polls show it—that there are probably a majority of Catholics who would be happy with what they perceive as justice and recognition—cultural recognition—within Northern Ireland, and economic things remaining the way they were. Now if in fact it turned out that economically it was better to live in the Republic of Ireland, you would then have a completely different ballgame. But for the most part—I mean most middle-class Catholics have begun to rise through the ranks. They've begun to get recognition in public lobbies. They've begun professionally to become equal, and the number of Catholics in the universities now studying to be doctors and lawyers is greater than that of Protestants. So that's moved significantly, and there's a lot more recognition in fact that things are not so bad for Catholics in Northern Ireland. But I also think there is a romantic attachment—a little like St. Augustine—"I want the dream but not yet." There's a romantic attachment to the idea but not necessarily the reality. Interestingly, more Protestants believe there's going to be a united Ireland than Catholics. I mean, you can tell it from their fear. Most Protestants at some level fear that within the next 15 years, 20 years, they are actually going to be pushed into a united Ireland. Most Catholics would be horrified at this thought.

## Humanity

Identity in Ulster is painted on gable-end walls, pounded on drums, and paraded in the streets. How do we account for the wide range of visual and expressive forms and behaviors? For purposes of analysis and description, I have made some effort to discuss these forms as distinct categories, but in practice that is not possible. Lambeg drums, orange lilies, flags, and banners may all be experienced individually on discrete occasions, but they are also experienced together in parades, during which murals provide a visual and material environment. I have developed the concept of assemblage as a means of approaching smaller and larger collections of artifacts in ways that approximate actual perceptions. Further, we have seen that even the most "innocent" of these forms, regardless of a user's intentions, is deeply implicated in the political life of Ulster and its internal struggles, which various residents interpret as territory, nationality, religion, and so forth. Like American gangs, which also use codes expressed in murals, and also clothing, to differentiate and stake territorial claims—and whose deaths are frequently public events and are marked by memorial walls and spontaneous shrines—in Ulster, ignoring the territorial claims of a wall painting or attempting to stop a parade can lead to death.

When I say that in Northern Ireland, art and politics are one, it is this relationship of the personal to the public, and the very real passions and consequences that inform these activities, forms, and traditions, that I have in mind. The spontaneous shrines express personal grief, and define personal relationships, in the most public way. Nor is there the same division between intellect and passion that there often is in the United States. The people of Northern Ireland do not mistake passion for intellect, but their passion is intellectually grounded. It is engaged. They are passionate in their intellect and intellectual in their passion. In the context of the political demonstration or the commemoration, for instance, even signs become more than simple messages. They become polysemic weapons with social force and power, part of a larger battle. The media, of course, play their own role, broadcasting and thus publicizing these photographs and signs nationally and internationally. Still, the signs are used first in the local context—in Belfast, for example, pictures of the Bloody Sunday victims were attached to a bridge, thus joining with the environment and becoming part of the culturescape, during the 1998 Bloody Sunday commemoration. Likewise, Sean Graham's has become a flashpoint in the Reroute Sectarian Marches movement. Signs such as "Stop the Bloody Murder" held at a peace demonstration in Dublin, "The IRA Killed My Wife," held by a man at the airport as Gerry Adams departed for the United States, or the pictures of the victims of Bloody Sunday paraded by their relatives at a commemoration of the event at Derry—all these public acts force politics to become personal. It is, I think, the most humane way people can engage in politics, to always consider the particular, personal ramifications of one's actions. And that, despite the horror of the bombs and the bullets, is what ultimately redeems the people of Northern Ireland. Throughout it all, they have never lost their capacity for outrage.

# Epilogue

That sense of outrage, of justice, of morality, has led to a delicate peace, as of the time of this writing in January 2001. The situation will remain volatile and fluid, of course, and like the residents of Ulster, Ireland, and the rest of the world, I cross my fingers and hope for the best. Perhaps a new idea will prevail—one that recognizes similarities rather than differences, acknowledging that the people of Ulster do indeed share a common, if conflicted, heritage, one distinct from either the Republic of Ireland or Great Britain. Their murals and parades, while they have been used differentially in the past, are manifestations of a tradition of public display the Ulster British and the Ulster Irish share with each other more than with any other group in the United Kingdom or Ireland.

# References Cited

Abrahams, Roger D. 1981. "Shouting Match at the Border: The Folklore of Display Events." In Richard Bauman and Roger D. Abrahams, eds., *"And Other Neighborly Names": Social Process and Cultural Image in Texas Folklore,* pp. 303–304. Austin: University of Texas Press.

Adamson, Ian. 1974. *The Cruthin.* Belfast: The Pretani Press.

Alford, Violet. 1959. "Rough Music." *Folklore* 70:505–518.

Anderson, Benedict R. 1983. *Imagined Communities: Reflections on the Origin and Spread of Nationalism.* London: Verso Editions/NLB.

Appadurai, Arjun. 1990. "Disjuncture and Difference in the Global Cultural Economy." *Theory, Culture, and Society* 7:295–310.

Aretxaga, Begona. 1997. *Shattering Silence: Women, Nationalism, and Political Subjectivity in Northern Ireland.* Princeton: Princeton University Press.

Austin, Joe. 1996. "Rewriting New York City." In *Connected: Engagements with Media,* ed. George E. Marcus, pp. 271–312. Chicago: University of Chicago Press.

Bakhtin, Mikhail. 1984 (1965). *Rabelais and His World.* Trans. Helene Iswolsky Bloomington, IN: Indiana University Press.

Ballard, Linda. 1991. *Tying the Knot.* Cultra: Ulster Folk and Transport Museum.

Ben-Amos, Dan. 1976. "Analytical Categories and Ethnic Genres." In *Folklore Genres,* ed. Dan Ben-Amos, pp. 215–242. Austin: University of Texas Press.

———. 1984. "The Seven Strands of Tradition: Varieties in Its Meaning in American Folklore Studies." *Journal of Folklore Research* 21:97–131.

Bryan, Dominic. 1998. "'Ireland's Very Own Jurassic Park': The Mass Media, Orange Parades, and the Discourse on Tradition." In *Symbols in Northern Ireland,* ed. Anthony D. Buckley. Belfast: The Institute of Irish Studies.

Buchanan, R.H. 1962. "Calendar Customs, Pt. 1." *Ulster Folklife* 8:15–34.

———. 1963. "Calendar Customs, Pt. 2." *Ulster Folklife* 9:61–79.

Buckley, Anthony D. and Kenneth Anderson. 1988. *Brotherhoods in Ireland.* Holywood, County Down: The Ulster Folk and Transport Museum.

Buckley, Anthony D. and Mary Catherine Kenney. 1995. *Negotiating Identity: Rhetoric, Metaphor, and Social Drama in Northern Ireland.* Washington, D.C.: Smithsonian Institution Press.

Burke, Peter. 1978. *Popular Culture in Early Modern Europe.* New York: New York University Press.

Charsley, Simon. 1987. "Interpretation and Custom: The Case of the Wedding Cake." *Man* 22:93–110.

Coffin, Tristam P. and Cohen, Henig. 1975. *Folklore from the Working Folk of America.* Garden City, N.Y.: Doubleday.

Corrette, Leigh. 1996. "You Can Kiss Me if You're Irish but You Can't Kiss Me if You're Queer: The St. Patrick's Day Parade." Paper presented at the meetings of the American Folklore Society.

Cressy, David. 1989. *Bonfires and Bells: National Memory and the Protestant Calendar in Elizabethan and Stuart England.* Berkeley: University of California Press.

Danaher, Kevin. 1972. *The Year in Ireland.* Cork: The Mercier Press.

Davis, Natalie Zemon. 1975. "The Reasons of Misrule." In *Society and Culture in Early Modern France,* ed. Natalie Zemon Davis, pp. 97–123. Stanford: Stanford University Press.

Davis, Susan G. 1986. *Parades and Power: Street Theater in Nineteenth-Century Philadelphia.* Philadelphia: Temple University Press.

de Certeau, Michel. 1984. *The Practice of Everyday Life.* Translated by Steven Rendall. Berkeley: University of California Press.

De Rosa, Ciro. 1998. "Playing Nationalism." In Buckley, Anthony D. (ed.) *Symbols in Northern Ireland.* Belfast: The Institute of Irish Studies.

Douglas, Mary. 1966. *Purity and Danger.* London: Routledge and Kegan Paul.

Drewal, Margaret Thompson. 1992. *Yoruba Ritual: Performers, Play, Agency.* Bloomington, Indiana: Indiana University Press.

Feldman, Allen. 1991. *Formations of Violence: The Narrative of the Body and Political Terror in Northern Ireland.* Chicago: The University of Chicago Press.

Frey, Nancy L. 1998. *Pilgrim Stories: On and Off the Road to Santiago.* Berkeley: University of California Press.

Gailey, Alan. 1977. "The Bonfire in North Irish Tradition." *Folklore* 88:3–28.

Gaudet, Marcia. 1998. "The World Downside Up: Mardi Gras at Carville." *Journal of American Folklore* 111:23–38.

Gillis, John R. 1994. *Commemorations: The Politics of National Identity.* Princeton: Princeton University Press.

Glassie, Henry. 1975. *All Silver and No Brass: An Irish Christmas Mumming.* Bloomington: Indiana University Press.

———. 1982. *Passing the Time in Ballymenone.* Philadelphia: University of Pennsylvania Press.

———. 1995. "Tradition." *Journal of American Folklore* 108:395–412.

Greenhalgh, Susanne. 1999. "Our Lady of Flowers: The Ambiguous Politics of Diana's Floral Revolution." In *Mourning Diana: Nation, Culture, and the Performance of Grief.* London and New York: Routledge.

Handelman, Don. 1990. *Models and Mirrors: Toward an Anthropology of Public Events.* Cambridge, England: Cambridge University Press.

Hass, Kristen. 1998. *Carried to the Wall: American Memory and the Vietnam Veterans Memorial.* Berkeley: University of California Press.

Hobsbawm, Erik and Terrance Ranger. 1983. *The Invention of Tradition.* Cambridge, England: Cambridge University Press.

Ives, Edward D. 1988. *George Magoon and the Down East Game War.* Urbana: The University of Illinois Press.

Jarman, Neil. 1997. *Material Conflicts: Parades and Visual Displays in Northern Ireland.* Oxford and New York: Berg.

Kear, Adrian and Deborah Lynn Steinberg, eds. 1999. *Mourning Diana: Nation, Culture, and the Performance of Grief.* London and New York: Routledge.

Kinser, Samuel. 1990. *Carnival, American Style: Mardi Gras at New Orleans and Mobile.* Chicago: The University of Chicago Press.

Ladurie, Emmanual Le Roy. 1979. *Carnival in Romans.* Translated by Mary Feeney. New York: George Braziller.

Laqueur, Thomas W. 1994. "Memory and Naming in the Great War." In *Commemorations: The Politics of National Identity,* ed. John R. Gillis, pp. 150–167. Princeton: Princeton University Press.

Lawrence, Andy. 1996. *The Thompsons.* Manchester, England: Granada Centre for Visual Anthropology.

Loftus, Belinda. 1986. "Matters of Life and Death: Protestant and Catholic Ways of Seeing Death in Northern Ireland." *Circa Art Magazine* 26:14–15.

———. 1988. "In Search of a Useful Theory." *Circa Art Magazine* 40:17–23, 65–66.

———. 1992. *Mirrors: William of Orange and Mother Ireland.* Dundrum, Northern Ireland: The Picture Press

———. 1994. *Mirrors: The Orange and the Green.* Dundrum, Northern Ireland: The Picture Press.

Moore, Sally and Barbara Myerhoff. 1977. *Secular Ritual.* Assen, Netherlands: Van Gorcum.

Muir, Edward. 1997. *Ritual in Early Modern Europe.* Cambridge, U.K.: Cambridge University Press.

Newall, Venetia. 1972. "Two English Fire Festivals in Relation to their Contemporary Social Setting." *Western Folklore* 31:244–74.

Newman, Simon P. 1997. *Parades and the Politics of the Street: Festive Culture in the Early American Republic.* Philadelphia: University of Pennsylvania Press.

Noyes, Dorothy. 1995. "Façade Performances: Public Face, Private Mask." *Southern Folklore* 52:91–96.

Peteet, Julie. 1996. "The Writing on the Walls: The Graffiti of the Intifada." *Cultural Anthropology* 11:139–159.

Roach, Joseph. 1996. *Cities of the Dead: Circum-Atlantic Performance.* New York: Columbia University Press.

Robinson, Philip S. 1994. *The Plantation of Ulster: British Settlement in an Irish Landscape, 1610–1670.* Belfast: The Ulster Historical Foundation.

Rolston, Bill. 1991. *Politics and Painting: Murals and Conflict in Northern Ireland.* London: Associated University Presses.

———. 1992. *Drawing Support: Murals in the North of Ireland.* Belfast: Beyond the Pale Publications.

Santino, Jack. 1983. "Halloween in America: Contemporary Customs and Performances." *Western Folklore* 62:1–20.

———. 1986. "The Folk Assemblage of Autumn: Tradition and Creativity in Halloween Folk Art." In *Folk Art and Art Worlds,* eds. John Michael Vlach and Simon Bronner. Ann Arbor, MI: UMI Research Press.

———. 1988. "The Tendency to Ritualize: The Living Celebrations Series as a Model for Cultural Presentation and Validation." In *The Conservation of Culture: Folklorists and the Public Sector,* ed. Burt Feintuch, pp. 118–131. Lexington, KY: The University Press of Kentucky.

———. 1992a. "'Not an Unimportant Failure': Rituals of Death and Politics in Northern Ireland." In *Displayed in Mortal Light,* ed. Michael McCaughan. Antrim: Antrim Arts Council.

———. 1992b. "Yellow Ribbons and Seasonal Flags: The Folk Assemblage of War." *Journal of American Folklore* 105:19–33.

———. 1994a. *All Around the Year: Holidays and Celebrations in American Life.* Urbana: University of Illinois Press.

———. 1994b. *Halloween and Other Festivals of Death and Life.* Knoxville: University of Tennessee Press.

———. 1996. "Light Up the Sky": Halloween Bonfires and Cultural Hegemony in Northern Ireland. *Western Folklore* 55:213–231.

———. 1998. *The Hallowed Eve: Dimensions of Culture in a Calendar Festival in Northern Ireland.* Lexington: The University Press of Kentucky.

Saxl, F. and Wittkower, R. 1948. *British Art and the Mediterranean.* Oxford: Oxford University Press.

St. George, Robert Blair. 1995. "Ritual House Assaults in Early New England." In *Fields of Folklore,* ed. Roger D. Abrahams. Bloomington, IN: Trickster Press.

Stoeltje, Beverly J. 1993. "Power and the Ritual Genres: American Rodeo." *Western Folklore* 52:135–156.

Thompson, E. P. 1963. "The Planting of the Liberty Tree." In *The Making of the English Working Class.* New York: Pantheon Books, 1963.

———. 1991. "Rough Music." In *Customs in Common,* ed. E. P. Thompson, pp. 467–538. New York: The New Press.

Travers, Len. 1997. *Celebrating the Fourth: Independence Day and the Rites of Nationalism in the Early Republic.* Amherst: University of Massachusetts Press.

Turner, Victor. 1967. "Betwixt and Between: The Liminal Period in *Rites de Passage.*" In *The Forest of Symbols,* ed. Victor Turner, pp. 93–111. Ithaca, N.Y.: Cornell University Press.

———. 1969. *The Ritual Process.* Ithaca, N.Y.: Cornell University Press.

———. 1974. "Social Dramas and Ritual Metaphors." In *Dramas, Fields, and Metaphors: Symbolic Action in Human Society.* ed. Victor Turner, pp. 23–59. Ithaca and London: Cornell University Press.

———. 1982. *From Ritual to Theatre: The Human Seriousness of Play.* New York: Performing Arts Journal Publications.

Waldstreicher, David. 1997. *In the Midst of Perpetual Fêtes: The Making of American Nationalism 1776–1820.* Chapel Hill, North Carolina: University of North Carolina Press.

Walter, Tony, ed. 1999. *The Mourning for Diana.* Oxford, UK: Berg.

Zimmerman, Thomas Anthony. 1995. *Roadside Memorials in Five South Central Kentucky Counties.* M.A. Thesis, Western Kentucky University. Bowling Green, KY.

# Index